NOV 0 6

EVOLUTION AND CHRISTIAN FAITH

Evolution *and* Christian Faith

❖

REFLECTIONS

of an

EVOLUTIONARY

BIOLOGIST

❖

Joan Roughgarden

⬤ **ISLAND**PRESS   *Washington · Covelo · London*

Island Press is a trademark of
The Center for Resource Economics.

Library of Congress Cataloging-in-Publication Data
Roughgarden, Joan.
Evolution and Christian faith : reflections of an
evolutionary biologist / by Joan Roughgarden. .
p. cm.
Includes index.
ISBN 1-59726-098-3 (cloth : alk. paper)
1. Evolution (Biology)—
Religious aspects—Christianity. I. Title.
BS659.R68 2006
231.7'652—dc22   2006009297

Printed on recycled, acid-free paper ✇

Manufactured in the United States of America

10 9 8 7 6 5 4 3 2 1

*To Persons of Faith
at St. Gregory of Nyssa in San Francisco
and Everywhere*

Contents

❋

Acknowledgments

✤

I thank Megory Anderson, founder of the Sacred Dying Foundation, for helping me to visualize how thoroughly faith could be integrated with science in this book, and Chuck Crumly and Blake Edwards for early brainstorming about the project. I thank the members of St. Gregory of Nyssa for their support, especially the rectors, Richard Fabian and Donald Schell, for their encouragement and concrete leads to relevant scripture, and the people who joined a roundtable discussion one evening of an early draft of the manuscript. I thank students at Stanford University for comments during seminar discussions of the material. I thank Patricia Jung of Loyola University–Chicago, William Countryman of The Church Divinity School of the Pacific, and Michael Rose of the University of California–Irvine for scholarly reviews. Finally, I thank my editors at Island Press, Barbara Dean and Jonathan Cobb, for their comments, and especially Barbara for the opportunity to discuss wording, pacing, and concept nearly every day. I'm deeply grateful to everyone whose encouragement has guided me in what has been an exciting opportunity for synthesis. Thank you all.

Evolution *and* Christian Faith

I

Science with Religion

OTH SIDES OUGHT TO BE PROPERLY taught," urged President George W. Bush when asked at a news conference in August 2005 about teaching evolution and its alternatives, such as intelligent design, in the public schools. I'm an evolutionary biologist and a Christian. Here's my perspective on what to teach about evolution and on how to understand today's collision between science and Christian faith.

It is 175 years since Darwin embarked on his 1831 voyage to South America in a ten-gun brig of the British navy called the *Beagle*. That voyage carried him to the Galápagos where he began his thinking about evolution. After all this time we can now say that some parts of the subject he started, called evolutionary biology, have been demonstrated as true. I believe the demonstrated parts should be taught in all our schools and that to not do so is like failing to teach that the earth is round. On the other hand, not all of evolutionary biology is as clear-cut as evolution's basic facts are, and some points are seriously problematic. This book clarifies what is truly established in evolutionary biology, indicates what aspects of evolutionary theory remain inadequate, and identifies some parts that are probably wrong. It will, I hope, give you a balanced view of the state of evolutionary biology today.

I'm a Christian and active in the Episcopal Church. My parents were Episcopal missionaries, and I grew up in the Philippine Islands and Indonesia. I attended church through high school, drifted away after college, and returned about ten years ago when I was facing personal trials. Since then, I've worked through the connection between my faith and my occupation as an evolutionary biologist.

Although we keep hearing of "sides" in a "debate" between evolution and religion, I don't think that way. After all, the famous "monkey" trial in Tennessee ended long ago, in 1925. This trial challenged the teaching of evolution in schools and is named the Scopes Trial after the defendant, John T. Scopes, who was a teacher. You may have seen this trial dramatized in the classic 1960 movie *Inherit the Wind,* with Spencer Tracy, and may recall that evolution came out the "winner." I ask, Why should we bother replaying that trial again and again today? I don't want to argue with other Christians. I want to share with them the fellowship and the love of Jesus.

Here we'll look at actual biblical passages and you can see for yourself whether any conflict exists with evolutionary biology. You may agree that the Bible doesn't have any necessary conflict with evolutionary biology, as many people already think. I think that, even more important, what evolutionary biologists are finding through their research and thinking actually promotes a Christian view of nature and of our human place in it. Thus, as Christians we don't have to simply stand aside and say, Well, science is about the material and religion about the spiritual, and ne'er the twain shall meet. Instead, we can rejoice as Christians in the ethical meaning behind what evolutionary biologists are increasingly finding. I've been exhilarated by this personal realization, and I hope you will be, too.

I've written this book for several reasons. First, I want to provide a short and succinct statement of what evolutionary biology is, what it says and what it doesn't say. Because many people get their idea of evolution secondhand, a lot of misunderstanding is creeping in and I think there's a need to set the record straight. Second, I want to discuss what the Bible actually says, for the Bible too is often misrepresented. Third, I want to investigate the relationship between evolutionary biology and the Bible. I want to do this because it pains me to see proponents of science and Christianity ridicule and hurt each other. I hope to get us talking constructively.

I've kept the book short—a read for a long plane trip, or a night or two's sitting to brief for a school board meeting, parent–teacher conference, or church group discussion, or for writing a term paper—any situation where you've got to come up to speed fast on today's incarnation of the controversy over teaching evolution.

I've been struck by how the "debate" over teaching evolution is not about plants and animals but about God and whether science somehow threatens one's belief in God. To analyze this threat, I'll discuss plants, animals, and God all together—something people rarely do. My specialty is lizards. When I lecture, I discuss what islands my lizards live on, what food they eat, what their colors are, and who their predators are. I never mention God. When anti-evolutionists lecture, they discuss God, what's in the Bible, and how they think the Bible should be interpreted. They never mention lizards. So, if we're going to get a dialogue going on evolution, we're going to have to start talking about plants, animals, and the Bible together in one place, as I do in this book.

In talking about plants, animals, and the Bible in the same paragraph, even in the same breath, are we fudging the separation between science and religion? Is this right? Maybe in some past world the spheres of science and religion were kept separate, but today the separation has disappeared—the polls tell us that we're past the point of no return.

A CBS News poll in November 2004, based on a nationwide telephone survey of 885 adults, showed that 65 percent of all Americans favored teaching creationism alongside evolution. A follow-up poll of 2,000 people by the Pew Forum on Religion and Public Life in July 2005 found that 64 percent were open to the idea of teaching creationism in addition to evolution. Both polls are practically identical and confirm each other. The polls also show large numbers, around 40 percent, in favor of replacing evolution with creationism in the science curriculum. When compared with these poll results, President Bush's position that both sides should be taught seems modest.

Similarly, Senator Bill Frist, a Tennessee Republican leader who is a medical doctor, also endorsed teaching both sides, saying such an approach "doesn't force any particular theory on anyone . . . I think in a pluralistic society that is the fairest way to go about education and training people for the future." Many scientists were surprised by Senator Frist's statement and dismissed it as an attempt to garner conservative Christian support for his anticipated presidential bid. One might anticipate that medical doctors, because of their substantial education in biology during premedical training, would be more supportive of teaching evolution than the general public is. But, in fact, many doctors are unsure of evolution, and Senator Frist's view is not extreme among them.

A poll in May 2004 of 1,472 physicians by the Louis Finkelstein Institute at the Jewish Theological Seminary together with HCD Research in Flemington, New Jersey, reported that 65 percent approved of teaching both evolution and creation in the schools. And 11 percent of the Catholic doctors, plus 35 percent of the Protestant doctors, believe that "God created humans exactly as they appear now." Acceptance of evolution is far from unanimous among Christian doctors.

The reason is not hard to find: evolution is increasingly crowded out of the premedical curriculum by topics like cell biology, physiology, and genetics, which are more pertinent to day-to-day medical practice than evolution is. The polls show that, as a group, doctors don't know any more about evolution than does the general public. And a majority of the public, including doctors, already wants science and religion taught side by side.

Although both President Bush and Senator Frist have voiced positions on evolution consistent with the polls, I believe it's fair to say that they have not assumed leadership in showing how evolution should be presented or in finding some way to bridge the perceived gap between evolutionary biology and Christian values. I hope that this book will help bridge that gap.

People often seem to want to talk more about religion than about science. I sense a deep desire to speak about God but not much desire to see another nature show. I believe this pent-up urge for talking about God is being met through the work-around of shoehorning God into a debate about teaching evolution. Why? Because it's inappropriate to discuss matters of religious faith in public debate, whereas it's acceptable to talk about the science curriculum. We talk about teaching evolution when we really want to talk about God, not plants and animals. So, let's not fool around. In this book, I speak openly and explicitly about God, not God camouflaged as science. Once the air is cleared, I hope we can return with rekindled interest to the plants and animals that motivated evolutionary theory in the first place.

Just as evolutionary theory has both rock-solid parts and squishy parts, some Christian teachings enjoy broad support while others are embraced by only a few sects. Evolutionary science threatens some Christians but not others. Why? When evolution seems to conflict with religious beliefs, could the problem be in the religious teaching, not the science teaching? A balanced look at teaching evolution necessitates looking seriously at *both* religion and science, which I do in this book.

People have asked me to be direct about my own definition of God and my own religious sentiments. Well, to me God is an experience, not an idea. I experience

God through human community. I believe Jesus lived and died for our sins. His teachings express a vision of love and inclusion. I am concerned not with "proving" whether God exists but with living a Christian life. Beyond that, I'd rather not make much of my own story, because this book is not about me. It's about burying the hatchet between science and religion, starting with the teaching of evolution.

Here's where I'm coming from. I've taught evolutionary biology and ecology for more than thirty years at the university level and have made a lifelong study of the subject. In contrast, I write as a simple parishioner about Christianity and the Bible. I wish I had the depth of experience with the Bible that I have with evolution, but I don't, and I hope you will forgive my mistakes. My faith tradition emphasizes a personal responsibility to engage the Bible and to tell others about it. Although I question my own competence to write about matters of Christian faith, I draw courage from the instruction in each Sunday's sermon about the need to speak up. Many of you will also have developed your own personal understanding of the Bible, and I hope that you will find my interpretations complementary to your own inquiry. I know this has been an important journey for me. I simply would not participate in evolutionary biology if I thought it somehow undercut Christian faith or was in any way immoral or destructive to our shared humanity. I think that people who actively engage in both evolutionary biology and Christianity should know in their hearts that they are doing the right thing.

For the biblical passages I use the King James Version. I've wanted to avoid digressing into the pros and cons of the dozens of different translations that are currently available and have chosen a version that is familiar, even though couched in Old English phrases. The points I'm making depend not on how a particular Hebrew or Greek word is translated but on the whole passage, the meaning of which comes through in any translation. Furthermore, I accept the Bible as is—this is the book on the table in front of us. I believe our first task is to see how the Bible, as currently given in a standard text, compares with the findings of evolutionary biology. You may wish to pursue further historical and textual analysis of the passages that I've singled out, but for now we have to start somewhere.

Finally, you might wonder why I am writing solely about Christian faith and its relationship to evolution. Well, I'm even less qualified to write about other faith traditions, or about atheism, than I am about Christianity. More important, the problem before us is specifically between evolution and Christianity. It is only Christians, not people of other faiths, who are challenging the teaching of evolution in our schools. If you are from other faith traditions, or are an atheist or agnostic, then I hope this book will still be of interest to you because of the status report it presents on the current state of evolutionary biology and because the relation of evolutionary biology to Christian teachings is a major issue of our time.

Here is the plan for the book. The early chapters present the facts of evolutionary biology. The middle chapters focus on how evolutionary biology explains those facts. The final chapters discuss current limitations to evolutionary theory as well as the challenges posed by present-day anti-evolutionists. Throughout, both scientific and Christian dimensions of the issues are discussed side by side.

I don't think we need more evidence for evolution. I think the presentation of evolution in our schools needs to respect the spiritual yearning of people that compels them to overlook the evidence we already have. I believe scientists need more sympathy and willingness to accommodate people of faith, to offer space for seeing a Christian vision of the world within evolutionary biology and not force people to accept a doctrine of universal selfishness as though established scientific fact.

2

Single Tree of Life

UST WHAT IS EVOLUTION? THE MAIN finding, originally suggested by Darwin in *The Voyage of the Beagle* in 1860, is that all life is related. All life belongs to one huge family tree. Some branches are the plants, other branches are animals, and subbranches of animals are the starfish, the insects, the mammals, and so forth, everything. The single tree of life is the basic fact of evolution. Not teaching that all life is related in one gigantic family tree is like not teaching that the Earth revolves around the sun.

You hear that evolution says we are descended from apes and monkeys. Sure, but that's not the point. *All* of life is related, not just humans with apes and monkeys. If you hug a tree, you're hugging a relative, a very distant relative of course, but a relative nonetheless. And we're not descended from apes and monkeys the way those species are now. Our relationship to monkeys and apes goes back in time to when our common ancestor lived—it's not that we were ever like today's monkeys or that they were like us, it's that we share ancestors in common.

How do we know this? The same way we tell whether people are related to one another. Children resemble their parents; they're not exact copies of course, but you can typically see the parents' eyes, nose, hair color, height, or another characteristic in their children. It's the same with animals, but you have to look beneath the skin, into the genes. Genes are substances deposited by parents into their eggs and sperm. After an egg and a sperm unite, these substances are "read" like a recipe so that the embryo can develop with the eyes, ears, nose, hair, height, and so forth that the parents have. Genes are made from a chemical called DNA, for short, although the full name, which nobody bothers to use, is several syllables long.

Everybody seems to know, even from courtroom television, that DNA can be used today for paternity analysis. Because children share the DNA of their parents, we can tell whether people are related by seeing if their DNA is the same. Well, this principle works across all living things. We share genes with all other living creatures—not just one or two genes but thousands of them. From 50 to 99 percent of all our genes are shared in common with other species, depending on how closely related we are to them. Today, the family tree of all life is being diagrammed from DNA data.

More evidence of evolution comes from fossils, the remains of ancient animals now embedded in rocks, but this is not as good as the DNA data because it's fragmentary. There are only so many quarries or exposed cliffs in which to dig up old bones and shells, and consequently the quantity of this data is always limited. But what there is shows the appearance through time of new forms coming from older forms. To me, what's interesting about fossils is not whether they prove the unity of all life, because the DNA data already do that, but that fossils offer a glimpse into what living conditions were like in the past. Digging fossils is the closest thing we have to a time machine. Fossils show that many species once lived that are no longer with us. This is a humbling realization, because some of these species were very widespread and looked totally successful, as though they would never disappear.

So if the major fact of evolution is that all life consists of one gigantic family tree, where's the controversy? Somehow, the idea is widespread that the single-tree-of-life discovery contradicts a literal reading of the Bible. Well, let's see. If you have a copy of the Bible nearby, turn to Genesis, Chapter 1.

After God created heaven and earth, light and dark, and the continents and oceans, he said, "Let the earth bring forth grass, the herb yielding seed, and the fruit tree." (Gen. 1:11) Then "God created the great whales, and every living creature that moveth, . . . and every winged fowl." (Gen. 1:20) Next, "God made the beast of the earth . . . and cattle . . . and every thing that creepeth." (Gen. 1:25) Finally, "God created man in his own image." (Gen. 1:27) Thus, God created the earth and populated it with plants and animals, and with people who in some way bear a resemblance to him.

Does Genesis 1 contradict evolution's tree of life? No. Genesis doesn't speak one way or the other about whether the plants and animals that God has placed on earth are related to one another. Surely God could create all of life as a single family tree if he wished.

I wonder why the impression is widespread that evolutionary biology somehow conflicts with the Bible. I've scoured the Bible for text relevant to the finding that all of life belongs to a common family tree but can't find passages pertaining directly to this particular fact. I've come to suspect that the illusion of conflict is caused by a poor presentation to the public of what evolutionary biology actually is. When many people find out what evolutionary biology actually says, its seeming threat to Christian faith dissolves.

One issue that may concern people is whether our sharing a family tree with other living creatures somehow demeans us. Does this relationship deny us a special place in the eyes of the Lord, and does it deny us our souls? No, evolution's finding pertains to our material connection with other living things.

We inherit our bodies from ancestors who were also the ancestors of other species. We do not inherit our souls from ancestors. Our resemblance in some way to God, that we are special in his eyes, is stated in Genesis 1:27: "So God created man in his own image, in the image of God created he him; male and female created he them." Even though God gave us this resemblance to him, he used material from the earth to build our bodies. Genesis 2:7 describes how God created Adam: "And the Lord God formed man of the dust of the ground, and breathed into his nostrils the breath of life; and man became a living soul." God gave us our souls even though building us from the substance of the earth. This is the very substance out of which he made all living things, like plants for instance. The Bible goes on to say, "And out of the ground made the Lord God to grow every tree that is pleasant to the sight, and good for food." (Gen. 2:9) Our possessing a soul is not challenged by the finding that all of life is united into one family tree, because the family tree is a material relationship, not a spiritual relationship.

One might go on to ponder when we humans acquired a "living soul" if we have descended from ancestors who presumably don't have souls. This puzzle is already solved. A person is connected materially to the sperm and egg that united to produce her or him. The sperm and the egg don't have souls, but depending on your faith tradition, sometime after they unite, or as they unite, the soul enters the embryo and it assumes the standing of a human person. The same solution appears true for the evolutionary descent of humans from ancestors who presumably didn't have souls. Sometime after the ancestors of humans split off on their own separate branch of the family tree, God breathed on them to give them living souls. When? I don't know, but my guess would be when our ancestors developed the capacity for discussing the difference between right and wrong. By then, if not before, our ancestors will have acquired souls and assumed the status of human beings.

Our material continuity with the rest of living creation is not a threat to Christian beliefs. Just the opposite. Our material connection with the rest of creation can be a source of joy. St. Paul's teachings in the New Testament explain the sacred significance of material connectedness. He uses the metaphor of a body consisting of many parts. Evolution's discovery of a single tree of life extends a Christian view of body and family beyond humans out to all of living creation.

St. Paul in his first letter to the Corinthians teaches that through sharing the bread of holy communion, Christians become one in body, "we being many are one bread, and one body: for we are all partakers of that one bread." (1 Cor. 10:17) Similarly, through sharing baptism, we become one in body, "For by one Spirit we are all baptized into one body, whether we be Jews or Gentiles, whether we be bond or free; and have been all made to drink into one Spirit." (1 Cor. 12:13).

Then St. Paul goes on to explain that the body has many interdependent parts and that no part can quit or be kicked out. "For the body is not one member but many. If the foot shall say, because I am not the hand, I am not of the body; is it therefore not of the body? ... But now hath God set the members every one of them in the body ... And the eye cannot say unto the hand, I have no need of thee ... And those members of the body, which we think to be less honourable upon these we bestow more abundant honour ... That there should be no schism in the body; but that the members should have the same care one for another. And whether one member suffer, all the members suffer with it; or one member be honoured, all the members rejoice with it. Now ye are the body of Christ." (1 Cor. 12:14–27)

This theme of a diverse community as one organic body, with each part different from one another yet essential to the whole, runs throughout St. Paul's teachings and also throughout evolutionary biology and ecology today. The fact that all of life is united into one family tree means that all of life belongs to one body and that all the varieties of plants and animals share as parts of this one body. Ecologists speak of an ecological community with each species having its "niche," or unique occupation, in the economy of nature. Ecologists also diagram the relationship among species as a "food web" that details how material travels within an ecological community from plants to leaf-eating insects, to lizards and birds who eat the insects, then to the biggest predators like hawks, and then back to the ground where their remains are recycled into a new generation of plants. From dust unto dust.

St. Paul's teaching that we are one body in Christ emphasizes the sacred significance to *material* unity in all of life. St. Paul's teachings are usually presented today in their historical context. St. Paul's letters served particular purposes, such as resolving squabbles among the early Christians living in Corinth, a Greek city between Athens and Sparta. But beyond such particularities, the vision projected by St. Paul's teachings uses the image of bodily continuity to emphasize the moral import behind material connectedness and therefore pertains to the moral import of the material connections that bind all of living creation together.

Similarly, St. Paul's concept of marriage and family life also expresses the idea of becoming one in body. In his letter to the Ephesians, St. Paul teaches that men should "love their wives as their own bodies . . . even as the Lord the church: For we are members of his body, of his flesh, and of his bones." (Eph. 5:28–30) Meanwhile the wife should honor her husband because he is part of her body too, "and the wife see that she reverence her husband." (Eph. 5:33) The fact that all of life is organically related as a single body in one family tree implies that St. Paul's teaching about how husband and wife should care for each other also applies to how mankind and living creation should care for each other.

The conventional Christian justification for conservation relies on "stewardship." Yes, stewardship is part of our moral responsibility, but not the whole picture. According to Genesis 1, God said to man and woman, "Be fruitful, and multiply, and replenish the earth, and subdue it: and have dominion over the fish of the sea, and over the fowl of the air, and over every living thing that moveth upon the earth." (Gen. 1:28)

The meaning of dominion is usually softened to mean stewardship, even though the verse also includes the word "subdue," which is hardly soft. For example, the section on Caring about Creation of the Episcopal Church's "Catechism of Creation" states, "'Dominion' does not mean 'domination,' but refers to the need for humans to exercise stewardship over the earth on God's behalf. In Genesis 2, the human beings are given the garden to tend and serve, symbolizing our obligation to care for creation."

Indeed, Genesis 2 does charge mankind with being stewards, and also students, of nature. "The Lord God planted a garden eastward in Eden; and there he put the man whom he had formed. . . . And the Lord God took the man, and put him into the garden of Eden to dress it and to keep it. . . . And out of the ground the Lord God formed every beast of the field, and every fowl of the air; and brought them unto Adam to see what he would call them: and whatsoever Adam called every living creature, that was the name thereof. And Adam gave names to all cattle, and to the fowl of the air, and to every beast of the field." (Gen. 2:8–20) In this way, mankind is charged with stewardship (dressing and keeping) and study (naming) of living nature.

Moreover, the Bible is clear that the responsibility for stewardship and study pertains to *all* of living nature, to lowly bugs and cockroaches as well as soaring eagles and majestic lions. Before the great flood, Noah put representatives of *all* species on the ark, not some privileged subset: "Of clean beasts, and of beasts that are not clean, and of fowls, and of every thing that creepeth upon the earth, There went in two and two unto Noah into the ark." (Gen. 7:8–9) And after the floodwaters had receded, here's what disembarked from the ark: "Bring forth with thee every living thing that is with thee, of all flesh, both of fowl, and of cattle, and of every creeping thing that creepeth upon the earth; that they may breed abundantly in the earth, and be fruitful, and multiply upon the earth." (Gen. 8:17) Especially important is that the list includes both "clean" and "unclean" species—that is, both edible and inedible species. A species is not valued because it is useful to mankind or approved by mankind. Instead, absolutely every species is intrinsically valuable—every kind of life, every species, every genotype of individual has equal worth in the eyes of God because all are part of his creation and sailed as passengers together through stormy weather on the great ark.

But to say that the moral principle of conservation in the Bible boils down to stewardship of nature is incomplete. Our relationship to nature is not merely one of benevolent boss, it is one of love, because we are one body with nature. We must be good husbandmen to living nature as husband is to wife. We must care for living nature as we care for ourselves, and nature gives us of its body when respected and cherished.

Is there then a conflict between the Bible and evolution? No. To the contrary, the discovery that all of life is one body through its union into one family tree extends St. Paul's teaching on Christian community to all of living creation. This finding is a source of joy, and I rejoice.

3

Species Change

HE SECOND FACT OF EVOLUTIONARY biology is that species change. The size, shape, and color of the animals and plants in a species are not cast in stone but change through time and space. This is very important—it shows the essential difference between a species of living things and of nonliving things. A chemical species, say, water, never changes. At sea level, it always boils and freezes at the same temperature, now and forever. Water is always colorless. There was never a time or place when pure water was red or green. But species of living things do change.

On these two facts hang all of evolutionary biology: one family tree unites all of life, and species change through time and place.

There's a lot to evolutionary biology today, but the single-tree-of-life and species-change discoveries are the bottom-line, take-home facts. I believe these two facts *must* be taught in any science curriculum today. Not teaching these will cripple the minds of children, as though asking them to find a place in modern society after being raised by wolves.

Again the question comes up about whether the Bible disagrees. Does the Bible say species can't change? Consult Genesis again. Nothing there prevents a species that God places on the earth from changing afterward. In Genesis 1:22, after God populated the earth with fish and fowl, the Bible says, "And God blessed them, saying, Be fruitful, and multiply, and fill the waters in the seas, and let fowl multiply in the earth." Well, absolutely nothing in these passages requires the fish and fowl to make exact copies of themselves as they multiply and cover the earth.

Where has the common belief come from that the Bible somehow rules out species from changing? If the Bible really said that, then there would indeed be a conflict with evolution. The passages from Genesis usually interpreted as having something to do with whether species can change involve the phrases "after his kind" or "after their kind." For example, Genesis 1:11–12 states, "And God said, Let the earth bring forth grass, the herb yielding seed, and the fruit tree yielding fruit after his kind, whose seed is in itself, upon the earth: and it was so. And the earth brought forth grass, and herb yielding seed after his kind, and the tree yielding fruit, whose seed was in itself, after his kind: and God saw that it was good." In this passage, God creates the grasses, herbs, and fruit trees to begin with and sets them to reproduce themselves with their own seeds. What the Bible says about whether species can change over time depends on how much one reads into "after his kind."

Some scholars have debated how accurately the phrase "after his kind" reflects the ancient Hebrew text. Instead of pursuing that approach, it's more relevant biologically to point out that the Bible in no way defines what sort of grouping "kind" refers to or what's involved in naming a "kind." Some Christian writers assume kind refers to a whole phylum, a major grouping like the chordates, mollusks, or crustaceans. These "high" categories describe a characteristic body plan—insects have shells with jointed appendages, whereas vertebrates have an interior skeleton with bones, and so forth. Instead, I think "kind" refers to "low" categories, like the varieties within a species, even particular genes within a species. For example, Genesis 1:25 speaks of God making the "cattle after their kind." All cattle belong to the same species because they can interbreed, and Genesis here is speaking of God populating the earth not only with species but also with the varieties within the species. Thus, God is creating genetic diversity from the very beginning.

Furthermore, the criteria for naming a kind don't

require that the organisms being classified remain static. When animals are introduced to a new habitat, they change over generations, but we still use the same name for them until they become different enough that a new name is introduced to avoid confusion. For example, the dingoes of Australia are wild dogs introduced from Asia. From a photograph they look like an ordinary dog we might have as a pet sleeping by the fireplace. Indeed, for many years they were classified as a variety of domesticated dog (*Canis familiaris*), but recently the name was changed to a variety of wolf to better reflect their temperament and natural habits (*Canis lupus*). Thus, the dog could change quite a bit over generations and still be called the same kind, a dog, until enough differences accumulate that some zoological dictionary writer throws in the towel and dreams up a new name.

It's clear that the Bible isn't making any statement one way or the other about whether species can change after they've been created. To the contrary, the reference to kinds of cattle implies that "kinds" actually can change. People bred cattle into different lines—some for beef, some for milk, and some to pull carts and wagons. Both Old and New Testaments often dignify the occupation of a husbandman who tends grapes and the shepherd who tends sheep. These are the farmers of old who made the cattle after their kind.

For example, Genesis 30:30–31:12 offers an extended

account of breeding livestock by Jacob, a servant of God. The scene is part of a long narrative spanning several chapters. The gist at this point is that Jacob, as recompense for previous injustices, makes a deal with his master, Laban, to keep for himself the cattle and goats that are speckled and spotted and the sheep that are brown. Any cattle and goats that are not speckled and spotted as well as sheep that are not brown will revert to Laban's ownership. "And it came to pass, whensoever the stronger cattle did conceive . . . the feebler were Laban's, and the stronger Jacob's." (Gen. 30:41–42) "And the angel of God spake" unto Jacob and said "Lift up now thine eyes, and see, all the rams which leap upon the cattle are . . . speckled . . . for I have seen all that Laban doeth unto thee." (Gen. 31:11–12) In this way, God reimbursed Jacob for the past injustices he had suffered by causing the speckled cattle he owned to be those that did most of the breeding, resulting in Jacob's part of the herd prospering while Laban's dwindled. God's hand molded the evolution of the livestock in Jacob's favor by determining which rams bred, the ones who "leap upon" their mates.

In this way, livestock, like our dogs, cats, and other pets, change in time through breeding. Evolutionary change is the natural counterpart of plant and animal breeding, but it goes on even when the farmer is asleep for the night—it doesn't need a farmer to select some animals for placing in a breeding pen or to carry pollen from some flowers to other flowers. When breeding takes place naturally, some individuals breed and others don't, all by themselves, which winds up causing species to change through time just as surely as when a farmer breeds cows. This evolutionary change is how the mystery of God's design for nature unfolds, how it becomes revealed to us in the fullness of time. (Rom. 16:25–26) I think the Bible is perfectly consistent with the two main facts of evolution—that all of life belongs to a common family tree and that species change over generations. Now that we've examined these facts and looked at the Bible, I hope you'll agree.

4

Taking the Bible Literally

OME OF YOU MAY BE SURPRISED that I've bothered to compare the two main facts of evolutionary biology, one tree of life and species change, with Genesis. Many Christians take the Bible as a general guide to living and find it a waste of time to compare the facts of evolution to the literal text of the Bible.

People often dismiss biblical literalists. In the nineteenth century, the Englishman Samuel Birley Rowbotham claimed the earth must be flat because the Bible uses the phrase "ends of the Earth" in several places. More recently, the late Charles Johnson of Lancaster, in the desert of Southern California, was head of the International Flat Earth Research Society, which also argued its case from biblical literalism. Positions such as these tend to undercut the legitimacy of biblical literalism as a theological stance.

In contrast, the theologian John MacArthur states the literalist case well in his book *The Battle for the Beginning: The Bible on Creation and the Fall of Adam*: "The starting point for Christianity is not Matthew 1:1 but Genesis 1:1. Tamper with the Book of Genesis and you undermine the very foundation of Christianity . . . If Genesis 1 is not accurate, then there's no way to be certain that the rest of Scripture tells the truth . . . If you reject the creation account in Genesis, you have no basis for believing the Bible at all. If you doubt or explain away the Bible's account of the six days of creation, where do you put the reins on your skepticism?"

I agree it's a problem to decide where to put the reins on skepticism. One has to choose somewhere, either the literal text or the facts. For me, it's the literal text, where I choose carefully, making sure I accord with the facts. Why? Because that's what I believe Jesus teaches.

Jesus consistently tried to get people to think for themselves, to not rely on fixed rules or literal text. His teaching is particularly clear on the do-and-don't rules that appear in Leviticus, one of the earliest books of the Old Testament. Leviticus is filled with detailed instructions for day-to-day life, including diagnosing and treating diseases, selecting acceptable food items, and proscribing punishments for adultery, complete with different punishments for a man sleeping with his brother's wife, his father's wife, his uncle's wife, and so forth. Concerning diet, for example, it's okay to eat fish, "whatsoever hath fins and scales in the waters, in the seas, and in the rivers, them shall ye eat." (Lev. 11:9) But mammals like seals presumably are not okay. "Whatsoever hath no fins nor scales in the waters, that shall be an abomination unto you . . . they shall not be eaten." (Lev. 11:12–13) And so on.

In Matthew, the first book of the New Testament, Jesus comes right out to reject this rule-based approach to living. Concerning diet he says, "Hear and understand: Not that which goeth into the mouth defileth a man; but that which cometh out of the mouth, this defileth a man." (Matt. 15:10–11) The disciples then told Jesus that some clergy were offended by this teaching. Jesus came right back and reiterated the point, "Do not ye yet understand, that whatsoever entereth in at the mouth goeth into the belly, and is cast out into the draught? But those things which proceed out of the mouth come forth from the heart; and they defile the man. For out of the heart proceed evil thoughts, murders, adulteries, fornications, thefts, false witness, blasphemies: These are the things which defile a man: but to eat with unwashen hands defileth not a man." (Matt. 15:17–20) Jesus could not have been clearer, even though the clergy at the time were angry with him for

saying so. The rules are secondary. The ethical principle is what matters most. If you eat the wrong food, it will soon pass through the body and be forgotten; but if you speak evil, it comes from the heart and is wrong. Jesus was perfectly willing to overrule the established teaching of clergy in his day who were missing the point even though those clergy could claim some textual support for their position from early scripture.

So, how do I choose what to understand from the Bible? I try to follow Jesus' example. Jesus did not arrive on the scene to say, "Here are some new rules to replace the old ones." He didn't say, "It's now okay to eat seal," as though updating a weight watcher's diet program by adding chocolate. No, Jesus broke from the mold altogether, and he taught through the use of parables, not rules. Jesus wants us to grasp his substantive point and then apply it to our present-day circumstances. I think Jesus would be aghast if he saw people trying to deny demonstrated facts in his name. We should not be alarmed when a fact conflicts with some literal text in the Bible because Jesus has already dealt with the problem of needing to refine the interpretation of older scripture in light of current knowledge and circumstance.

The earth wasn't made in six of our twenty-four-hour days. It's taken a long time to get where we now are. However, evolutionary biology doesn't have responsibility for determining how old the earth is.

That's geology. We biologists accept geological data and work with geologists. Our data splice nicely into geological data and often provide useful clues to how the earth works. For example, our data on where plants and animals live helped to uncover the fact that continents move—they're floating on a sea of molten lava. But that discovery is not part of biological evolution itself, although an evolution course would probably mention continental drift in addition to biological topics.

Concerning biological evolution itself, the key facts are that all life belongs to one family tree and that species change through time. These facts of biological evolution don't conflict with the literal text of the Bible. The chronology of biological evolution, which derives from geology, does conflict with the six-day time frame presented in Genesis 1. Yet the geologic fact that the earth is very old, and wasn't made in six days, is so clear that I'll follow Jesus' example of how to deal with ancient scripture and side with the evidence.

An objection I hear to understanding scripture in the context of today's scientific knowledge is that this is an unprincipled compromise. The approach I take, called "theistic evolution" by creationists, is the theological position that God creates the world according to a plan that continues to unfold, in the fullness of time, through the natural processes that science studies. This point of view has surely occurred independently to a great many people over the years who have wondered about the supposed conflict between science and religion.

Yet I hear people drawn to textual literalism disparage theistic evolution as nothing more than a middle-of-the-road convenience, a "trying to have it both ways." Against such presumed moral sloppiness, the passage in I Kings 18:21 is cited, in which Elijah chastises people for not taking a stand on behalf of their God: "And Elijah came unto all the people, and said, How long halt ye between two opinions? If the Lord be God, follow him: but if Baal, then follow him."

For me the position that God created the world, and continues to create it, through natural processes is not a compromise. God created those very processes as part of nature, so why should he jump outside of them? The Bible offers no grounds to believe that God would rather work in ways inexplicable as natural processes, even though he could do so if he wished. Just the opposite—jumping outside natural processes would imply something inadequate in the ability of natural processes to carry out his design. When God saw his creation, he said "it was good." (Gen. 1:12) You can't be consistent with the Bible on this point and yet also say that natural processes are inadequate to achieve God's design for his creation.

Let us turn now to what the natural processes are that bring about evolutionary change in living nature.

5

How Change Happens

OW WE GET TO THE NITTY-GRITTY of evolutionary biology today: *how* species change. Now we move beyond fact to theory. By "theory" I mean what explains the facts. Personally, I think evolutionary biologists have got right about 90 percent of what's going on. The rest still isn't clear, so research in this area is often fiercely debated. Here's the 90 percent that's correct; we'll get to the other 10 percent later.

Natural breeding is the main way species change through time, a scientific idea that goes back to Darwin's *Origin of the Species* in 1859. As mentioned earlier, species do change simply because in nature some individuals breed and others don't, all by themselves, so species come increasingly to resemble those that do breed and not those that don't.

In the 1800s Darwin used the phrase "natural selec-

tion" to describe this phenomenon. Darwin was draw-
ing a distinction between "artificial selection," which is
what farmers do, and "natural selection," which is what
occurs by itself in nature. The word "selection," however, has become a more general word for us, used for
everything from shopping at a store to picking furni-
ture to choosing music, whereas "breeding" still refers
only to what does or doesn't produce offspring. So I'll
use the phrase "natural breeding" instead of "natural
selection" to capture Darwin's original meaning in
today's language.

The phrase "natural breeding" also de-emphasizes
the often-used phrase "survival of the fittest." Evolution is not primarily about survival, but about what
breeds. Obviously, there's no breeding without survival, but survival alone is just a precondition for
breeding, and it's the successful breeding itself that
matters. Also, the word "fittest" tends to overempha-
size physical fitness, as though evolutionary theory
were an advertisement for daily workouts in the neigh-
borhood gym. "Fit" also means "fitting into" one's sur-
roundings, and survival of the fittest refers to how well
one meshes with one's environment, both social and
physical, as much as it does to physical vigor.

Although Darwin is credited with introducing natural breeding to biology as a scientific principle, the
idea behind natural breeding must have been known
since the dawning of agriculture.

Indeed, a metaphor for natural breeding is often used even in the Bible. In St. John's Gospel, Jesus describes a pastoral relationship between himself and his Father, and then between himself and his disciples, using what amounts to today's evolutionary terminology. Jesus says, "I am the true vine, and my Father is the husbandman. Every branch in me that beareth not fruit he taketh away: and every branch that beareth fruit, he purgeth it, that it may bring forth more fruit . . . I am the vine, ye are the branches: He that abideth in me, and I in him, the same bringeth forth much fruit: for without me ye can do nothing. If a man abide not in me, he is cast forth as a branch, and is withered." (St. John 15:1–6) That's natural selection, natural husbandry, there in the Bible, not as a one-liner but in an extended passage from Jesus himself.

All in all then, the way natural breeding works is obvious, has been known through history, and is part of the Christian tradition. So why is there any controversy? It's because there are two components to natural breeding, one that is clear—the natural husbandry part—and the other not so clear, which is the controversial part.

Suppose you're a chicken farmer trying to breed chickens to lay eggs that have a deeper yellow yolk than at present because it looks so good on toast. Well, you can check the eggs from the hens you have and see which hens lay the eggs with the yellowest yolk. You can let these hens mate with a rooster and raise their chicks to become the next generation. When these chicks grow up, they will, as a group, make eggs that have a deeper yellow yolk than their parents' generation did.

But wait. What if all the chickens made eggs with the same amount of yellow to begin with? Then there's no chicken that's any better at making yellow yolk than the others. So what can you do? Nothing. You're stuck. If you want to breed chickens, you not only have to be willing to locate which hens make the eggs you want and allow only those to breed, you also have to have some different kinds of chickens to start out with. If all the chickens are the same at the beginning, you're out of luck.

The same is true in natural breeding. If a species consists of individuals that are all the same to start with, it won't change. A species changes only if there are different kinds of individuals in it *and* those individuals breed at different rates. This point is basic.

Now we get to the controversy. Darwin was asked, "Why are any individuals different to begin with, why don't all chickens have the same color yolk to start with?" He couldn't answer. When Darwin wrote, genes hadn't been discovered. On the prevailing theories of inheritance in his day, any variation among individuals would simply dissolve over time, leading all the animals in a species to automatically be the same.

Thus, on one hand, everyone agreed that certain

shapes, colors, and behaviors would work best for an animal in any particular environment, and so yes, certain individuals would wind up breeding more than others, making the species eventually resemble those successful breeders. But on the other hand, no one could agree on why individuals were different to start with.

In today's evolutionary theory, these issues are called "source of selection" and "source of variation." The environment is the source of the selection—that's where certain traits emerge as more effective than others. But where is the source of variation?

When genes were discovered through the work of the Augustinian monk Gregor Mendel in 1865, the mystery seemed solved. According to Mendel, the reason that variation doesn't just dissolve away is because it is produced by genes that are transmitted from parents to children when the sperm combines with the egg to produce an embryo. The sperm contains the father's genes and the egg the mother's genes, and the child inherits genes from both parents when the sperm and egg unite. I'm sure you just take this fact for granted, but not too long ago it was big news.

The discovery of genes allowed scientists to fix up Darwin's account of natural breeding by saying that the source of the variation lies with the genes. Individuals are different from one another because they have different genes.

Unfortunately, the discovery of genes, although helpful, doesn't completely settle the matter. Granted, variation doesn't just dissolve away as originally feared because genes don't dissolve, but hmm . . . the discovery of genes just kicks the question of where variation comes from back to another level—why are there different genes?

6

Random Mutation

FTER DARWIN, A NEW CROWD OF
evolutionary biologists added genes
to Darwin's theory. Darwin plus genes
is called "neo-Darwinism." Leaders of
neo-Darwinism included the British
biologists Ronald Fisher and J. B. S. Haldane and the
American Sewall Wright, who all wrote primarily
from the 1920s to the 1940s. They set the stage for
evolutionary theory as we know it today. The neo-
Darwinists suggested that the genetic variation needed
for evolution by natural breeding came from mutation.

When a gene is copied and placed in the egg or
sperm for inclusion in an embryo, the copy may not be
exact. Any difference between the copy and the original is called a mutation. As we all know, in the real
world, copies are not always 100 percent identical to

the original. Gene-copying is pretty accurate; only about one in a million copies differs from the original—this is called the mutation rate.

Neo-Darwinists theorized that genetic differences between individuals are inevitable because of mutation. So, even if all your chickens happen to start out with the same yellow color of yolk, be patient, because eventually a chicken will turn up with a deeper color of yolk than the rest. When it does, then go for it! Use that hen for subsequent breeding and go on to produce a stock of chickens with a deeper yellow yolk than the original stock.

Fine so far, but just what are these mutations like, these differences between copy and original? It's tempting to say that any difference between copy and original is automatically an undesirable error. But suppose we're using a Xerox machine to copy a photograph with a boring blue sky. Perhaps some ominous dark gray clouds would turn the photograph from a ho-hum snapshot to an arresting dramatic landscape. Well, we know from our experience with office copy machines that a dirty roller leaves telltale smudges. Perhaps we can produce smudges in the copy by rubbing some ink on the roller. Then these smudges might give us just the dark gray clouds we need to fill an otherwise boring sky. If a mutation hurts the organism it turns up in, it's called deleterious; if it helps the organism it's in, it's called favorable. A smudge that spiffs up a boring snapshot with arresting dark gray storm clouds is a favorable mutation.

According to the neo-Darwinists, evolutionary progress awaits the appearance of favorable mutations. When they turn up, voilà, natural breeding takes off from there, resulting in evolution. This theory then invites the further question, Can we somehow steer evolution by determining what mutations occur before natural breeding takes over to further the evolutionary process?

In particular, it would be nice if the mutation process would produce the genes most useful later on in breeding. It would be great to shorten the wait for the first chicken with eggs with deep yellow yolks if we could somehow get the mutation process to produce this kind of chicken on an accelerated schedule. Once we get this chicken, then we can quickly make the whole stock evolve eggs with deeper yellow yolks through selective breeding.

Well, the bottom line is that we can't do it. Evolutionary biologists can't figure out how to put just the right ink on the roller to get just the right smudge on the copy to yield an arresting photograph from a boring original. We can't steer the mutation process to give us the genes we dream of having. We have to wait it out for desirable mutations to turn up on their own accord. This is called "random mutation." It means that the mutation process is independent of the subsequent breeding process—the source of variation is independent of the source of selection.

Just as the idea behind natural breeding must have been known since the dawn of agriculture, the idea behind random mutation must also have been known since the dawn of agriculture. And just as Jesus employed the natural breeding metaphor in his teaching, he also employed the random mutation metaphor in a parable.

In Matthew 13, Jesus speaks about the receptiveness of people to his message. He introduces the metaphor of a farmer (sower) casting small mustard seeds across the ground, some of which land on rock, others on poor soil, and still others on good soil. Jesus says, "Behold, a sower went forth to sow; And when he sowed, some seeds fell by the way side, and the fowls came and devoured them up: Some fell upon stony places, where they had not much earth: and forthwith they sprung up, because they had no deepness of earth: And when the sun was up, they were scorched; and because they had no root, they withered away. And some fell among thorns; and the thorns sprung up, and choked them: But other fell into good ground, and brought forth fruit, some an hundredfold, some sixtyfold, some thirtyfold. Who hath ears to hear, let him hear." (Matt. 13:3–9)

Jesus goes on to explain how his teachings are like

the mustard seeds. When his teachings fall on ears that do not hear, nothing permanent and worthwhile comes of it. But when his teachings fall on receptive ears like "seed into the good ground," the result "beareth fruit, and bringeth forth, some an hundredfold, some sixty, some thirty." (Matt. 13:23)

Random mutation is the same idea. You can think of random mutations as mustard seeds of DNA. If the soil is conducive to the seed's growth, if the body is receptive to the mutated gene, then the body grows and produces fruit a hundredfold. In the parable, God doesn't direct the seeds individually to good soil, nor does Jesus direct his teachings only to ears that can hear. The mustard seeds are broadcast, and the teachings spread, to all places and people, and then we discover what happens from there. The same is true of mutation. A variety of differences from the original are generated during the copying process, and then we see what natural breeding does. The mutations that agree with the body they find themselves in wind up reproducing themselves a hundredfold as the body containing them yields abundant fruit.

Unfortunately, the phrase "random mutation" has upset anti-evolutionists over the years. They apparently think this means biologists are claiming that evolution has no direction. No. Biologists are not claiming that. If they're claiming anything at all, it's that the direction to evolution comes from the breeding side of the evolutionary process, not the mutation-generating side.

What then is God's role in evolution? It depends on whether you view God as highly involved in everyday, minute-by-minute events or as setting the stage and then letting the plot unfold on its own.

Take the coin toss at the start of a football game. The coin could come up heads or tails. If it comes up heads, do you think God made it so, that God has a hand, so to speak, in each toss of a coin, giving your team the initial advantage? Or do you think God made people with the overall capacity to decide their disputes through coin-tossing and then takes no role in how a particular coin toss works out?

If you have a hands-on view of God, then God can steer evolution by controlling exactly which mutations occur and when, as well as by controlling which plants and animals get a chance to breed. If you have a hands-off view of God, then God doesn't intrude into each mutation but provides the circumstances in which some animals breed and others don't. A hands-on God steers evolution at both the mutation stage and the natural breeding stage, whereas a hands-off God steers evolution only at the natural breeding stage. A hands-on God reveals the design of creation by serving both as natural husbandman to direct who does and doesn't breed and as natural mechanic to direct what mutations occur. A hands-off God reveals the design of creation by serving as natural husbandman to direct breeding while leaving the mechanics of inheritance alone.

Where do biologists weigh in on this? We have no idea whether a particular coin toss will come up heads or tails. All we can give is the odds. Ditto for mutation—all we can give is the odds. For us, saying that mutation is random is a professionally acceptable way to admit ignorance. For us, saying that mutation is random has nothing to do with whether evolutionary change is fulfilling God's purpose or unfolding according to God's design. What then *do* evolutionists think about purpose and direction in evolution?

7

Evolution's Direction

LONG AND SOLID TRADITION TES-
tifies to biologists' search for direction
in evolution. Many, maybe most, evolutionary biologists *do* see evolution as
having a direction under the guidance
of natural breeding, even though the mutation-generating piece within the evolutionary process is random.

In 1954 the evolutionary biologist Theodosius Dobzhansky published an influential article entitled "Evolution as a Creative Process." He pointed out that the
outcome of natural breeding led to marvelous improvements in the breed and offered his experimental
results of natural breeding of fruit flies, his favorite
organism. The natural breeding occurred in his laboratory within little vials that housed hundreds of flies. He
did no selecting of which fly was, or wasn't, allowed to
breed; he just put a bunch of flies in the vial, let them

breed on their own for many generations, and watched what happened. The stock gradually became more productive as evolution progressed.

The idea that natural breeding leads to an improvement of the stock in terms of both ability to survive and number of offspring successfully reared runs throughout evolutionary theory. The idea has been explored mathematically in a continuous professional literature that stretches back to the 1920s.

Two early theoretical evolutionists, Ronald Fisher in the United Kingdom and Sewall Wright in the United States, developed mathematical theorems that showed how a species improves through time under the action of natural selection. These theorems all go by the rather pompous name of the Fundamental Theorem of Natural Selection. Actually, there are several variants of these theorems, and all show that in any before–after comparison of evolutionary change, a population of organisms winds up with individuals that are better able to breed than they were to begin with under the conditions in which the evolution is taking place.

Some of my own work in the 1970s explored the ecological impact of the continual evolutionary improvement envisioned by these venerable theorems. I showed that the outcome of evolution was a higher population size than before. I also investigated the impact on an ecosystem of evolution taking place within all the species living together. If species are competing with each other, then evolution in one species that improves it has the indirect effect of hurting its competitor. On the other hand, if a species is helping another species, then evolutionary improvements in it indirectly help other species too, a sort of rising-tide-lifts-all-boats result. The overall effect on an entire ecosystem of the evolution occurring within each of its species thus depends on the arrangement of interdependencies among the species.

Biologists skeptical of the idea that evolution has a direction challenged these theorems by trying to show that they may not always apply. Because mathematical theorems are always derived from specific assumptions, if one changes the assumptions enough, the result can be changed. So, paralleling the literature that evolution has a direction was a rival literature trying to debunk the idea of progressive evolution by changing the assumptions in the mathematical derivations. Anti-directionists seized on examples of nasty but successful genes, such as where one gene pushes out another when getting into the sperm and egg. The pushy gene then evolves instead of the polite gene. This is called "segregation distortion," and as it spreads, it hurts the organisms' capability to breed. So, sure, bad things can happen. Is this type of outcome very common, or does it last for very long? I doubt it. Evolutionists differ on how to evaluate the outcome of this challenge. I think the progressive-improvement position has the most merit, but it took some hits. Others might score the outcome differently.

None of this matters a whole lot—just a bunch of academics arguing, as they always do. The important point here is that evolutionary biologists are not of one mind that evolution is directionless, and some, including me, think it does usually have a direction of progressive adaptation to environmental conditions.

The direction to evolution that some biologists envision is not necessarily associated with a Christian or even a spiritual perspective. Some biologists, like Dobzhansky, wrote about the direction to evolution with a sense of spirituality. Others see the direction to evolution merely as a possible fact, without any spiritual or religious interpretation at all.

Thus evolution is not automatically in opposition to religion concerning a direction for evolutionary change. This allows much scope within evolutionary biology for envisioning that evolutionary change is fulfilling God's purpose, making nature unfold according to God's design. Whether this is actually happening is a religious belief that is neither supported nor opposed by evolutionary science. I recognize that many of my colleagues have no position on this question. For myself, I'm comfortable feeling that evolution by natural breeding is revealing God's design for nature in the fullness of time.

8

Roman Catholic Position

 ATTENDED AN INTERNATIONAL conference on evolution held in the Galápagos in June 2005 that was organized by the Universidad San Francisco de Quito in Ecuador. There the distinguished evolutionary biologist Antonio Lazcano of the Universidad Nacional Autónoma de México pointed out that the anti-evolution movement so vocal and powerful in the United States was absent in Latin America, a region with extensive membership in the Roman Catholic Church. Christian-based opposition to teaching evolution in the schools is primarily a U.S. phenomenon. "Why?" he asked. The reason he suggested is that the Catholic Church has generally not objected to evolution, as have fundamentalist Christians in the United States. Thus, he suggested, Christianity as such does not stand in opposition to teaching evolution, only some particular denominations do.

Indeed, the Roman Catholic Church has taken a nuanced and rather hands-off stance with regard to evolution, perhaps to avoid repeating its previous difficulties with Copernicus and Galileo. Copernicus was a mathematician who lived from 1473 to 1543 in Poland. Toward the end of his life he published a book proposing that the earth revolved around the sun, opposite to the classic belief endorsed by the Roman Catholic Church that the sun revolved around the earth. Galileo, an astronomer who lived in Italy from 1564 to 1642, then provided data using the telescopes he made himself to confirm Copernicus's theory. From 1616 until his death Galileo was constantly challenged and often imprisoned by the inquisition of the Roman Catholic Church for his conclusions. He was convicted of heresy. Finally, in 1992, Pope John Paul II admitted that errors had been made by the theological advisers in the persecution of Galileo.

In 1996, Pope John Paul II wrote a letter about evolution and Catholic doctrine. He began by recalling an official ruling (encyclical) made in 1950 by his predecessor Pope Pius XII, who "had already stated that there was not opposition between evolution and the doctrine of faith about man and his vocation, on condition."

The first condition of Pius XII was that the Church

not be thought of as actually endorsing evolution. As John Paul II phrased it, the 1950 Encyclical "considered the doctrine of 'evolutionism' a serious hypothesis worthy of investigation" but that "this opinion should not be adopted as though it were a certain, proven doctrine." John Paul II went on to say, "Today, almost half a century after the publication of the Encyclical, fresh knowledge has led to the recognition that evolution is more than a hypothesis," and he acknowledged the "series of discoveries" that have led to progressive acceptance "by researchers." In this way Pope John Paul II removed the first condition required by the 1950 Encyclical by acknowledging the increasing strength of data supporting evolution.

John Paul II reinforced the second condition in the 1950 Encyclical. He wrote, "Pius XII stressed this essential point: if the human body takes its origin from pre-existent living matter the spiritual soul is immediately created by God." He continued, "Theories of evolution which . . . consider the mind as emerging from the forces of living matter, are incompatible with the truth about man. Nor are they able to ground the dignity of the person." Evolutionary science is limited to studying the origin of the body, not the soul or the basis for human dignity. John Paul II says of the "sciences of observation" that "The moment of transition into the spiritual cannot be the object of this kind of observation."

I think John Paul II's 1996 letter is beautifully written, its tone loving and conciliatory. It clearly restricts science to the material and reserves the spiritual for God.

In July 2005, Cardinal Christoph Schönborn, the Roman Catholic archbishop of Vienna, editor of the 1992 Catechism of the Catholic Church, and who has, according to the religious Web site Beliefnet, been mentioned as a possible future pope, wrote an editorial in the *New York Times* dismissing John Paul II's letter as "vague and unimportant." He claimed that John Paul II didn't define what evolution was, and he went on to say that "evolution in the neo-Darwinian sense—an unguided, unplanned process of random variation and natural selection—is not . . . true."

Cardinal Schönborn refers to biologists as "defenders of neo-Darwinian dogma" and claims that "neo-Darwinists recently have sought to portray our new pope, Benedict XVI, as a satisfied evolutionist." Although one would have welcomed a more temperate statement, Schönborn's editorial is before us. Where is it right and where is it wrong?

A positive aspect is that the cardinal agrees that "evolution in the sense of common ancestry might be true." Teaching the main facts of evolution, one tree of life and species change, is reaffirmed as consistent with Roman Catholic doctrine.

Another constructive point is that Cardinal Schönborn brings to general attention some less-cited but relevant writings. Pope John Paul II stated in 1985 that "The evolution of living beings . . . presents an internal finality [purpose] which arouses admiration . . . [and] obliges one to suppose a Mind which is its inventor, its creator." A 2004 document of the International Theological Commission cautioned that John Paul's 1996 "letter cannot be read as a blanket approbation of all theories of evolution, including those of a neo-Darwinian provenance which explicitly deny to divine providence any truly causal role in the development of life in the universe" and continues with "An unguided evolutionary process—one that falls outside the bounds of divine providence—simply cannot exist." These additional writings help fill out the Roman Catholic position on evolution.

Now to the problems. Cardinal Schönborn confuses random mutation with evolution as a whole. Neither evolutionary biology nor neo-Darwinism specifically asserts that evolution overall is random, directionless, or unguided. As we saw in chapter 7, evolution may very well have a direction and be guided by the hand of God, even if the mutation process is random. The guidance is exerted at the natural breeding stage, not the mutation stage, of the evolutionary process.

Cardinal Schönborn presumes his opinion is fact. He writes that "one can clearly discern purpose and design in the natural world." However, for "purpose and design" to be considered facts of nature, some equipment would be needed or statistical test devised to demonstrate it. Otherwise, this claim is opinion, not fact.

Cardinal Schönborn asserts that one can prove the existence of God with data from nature. He writes, "The existence of God the Creator can be known with certainty through his works and by the light of human reason." Do you really think so? There have been thousands of years in which to discover God through force of data and pure intellect. If this were how to discover God, wouldn't we now know he exists as surely as we know the earth is round? Instead, I suggest the truth of God and his love is apprehended subjectively, and faith in God means acknowledging and dignifying our emotional being.

The controversy started by Cardinal Schönborn has now produced clarifications from him as well as from other high-ranking Roman Catholic clergy.

In January 2006, an interviewer from Beliefnet included on Cardinal Schönborn's Web site summarized the cardinal's position as follows: Cardinal Schönborn's "main concern . . . was not to denigrate evolution as a natural process but to criticize atheistic materialism as the dominant philosophy of today's secular societies." While the cardinal "believes that God is the intelligent designer of the universe, his position on evolution springs from a philosophical rather than a scientific standpoint."

These clarifications are helpful but do not go quite far enough. Cardinal Schönborn's own statements continue to misunderstand what evolutionary theory says. He directly states in the interview, "The argument that the whole complexity of life can be explained as mere random process is unreasonable in my opinion." Once again, evolutionary theory does not assert that the complexity of life is a "mere random process." It asserts only that the mutation process is random, not that evolution as a whole lacks direction.

In the January 2006 edition of the online Roman Catholic magazine *First Things,* Cardinal Schönborn claims that "my argument was based on careful examination of the evidence of everyday experience; in other words, on philosophy . . . my argument was superior to a scientific argument since it was based on more certain and enduring truths and principles."

Cardinal Schönborn cites scripture as the basis for the superiority of philosophy. He writes, "The mind's ability to grasp the order and design in nature . . . precedes faith, as Romans 1:19–20 makes clear . . . The natural world is nothing less than a mediation between minds: the unlimited mind of the Creator and our limited human minds." The cardinal is staking out a position in ontology, the branch of philosophy concerned with the meaning of reality—reality serves as a conduit between human minds and God's mind. He bases this approach to ontology on his interpretation of scripture.

In his letter to the Romans (1:20), St. Paul writes, "For the invisible things of him from the creation of the world are clearly seen, being understood by the things that are made." The point is that God becomes known through the study of his works, that is, nature.

On the whole, though, the Bible does not place as much confidence in human reason as Cardinal Schönborn evidently does. The entire book of Job offers a discourse on the limits to human reason, which would include philosophy as well as science. The book of Job records the conversations between God and Job, a wealthy man of the Old Testament, and offers the telling passage, "Great men are not always wise: neither do the aged understand judgment." (Job 32:9) In a later chapter God cross-examines Job, saying "answer thou me. Where wast thou when I laid the foundations of the earth? Declare, if thou hast understanding." (Job 38:3–4) And God continues his cross-examination for more than sixty more verses in the Bible, asking Job about all aspects of the physical and biological world. Job cannot answer and eventually says, "I uttered that I understood not; things too wonderful for me, which I knew not" (Job 42:3), and he repents. These passages offer a counterpoint to the trust in pure reason that Cardinal Schönborn apparently advocates.

A discourse on the philosophy of reality takes us far from the issue before us, which is what to teach about evolution in our schools. Indeed, Cardinal Schönborn is starting to distance himself from that issue altogether. Returning to the interview on Beliefnet, the interviewer reported that Cardinal Schönborn declined to take a position on a recent court case in the United States that found intelligent design not to be science, "saying he was not qualified to comment on American legal issues." Nonetheless, in the interview he was asked, "Does God belong in biology class?" Cardinal Schönborn replied, "The question of the Creator belongs in religion class. The question of the 'intelligent project of the cosmos,' as the pope put it, naturally belongs in with science." As we've seen, however, this boils down to whether there's a place in science class for a philosophical discussion of what "reality" means. If so, what will we do in philosophy class?

Other Roman Catholic clergy are beginning to speak up as well. The *New York Times* reported in November 2005 an interesting position from Rev. George Coyne, the Jesuit director of the Vatican Observatory, one of the world's oldest astronomical research institutions, based in the papal summer residence at Castel Gandolfo south of Rome.

Father Coyne is quoted as saying that "intelligent design isn't science and doesn't belong in science classrooms" and that "placing intelligent design theory alongside that of evolution in school programs was 'wrong' and was akin to mixing apples with oranges." He went on to say, "Intelligent design isn't science even though it pretends to be." Instead, "If you want to teach it in schools, intelligent design should be taught when religion or cultural history is taught, not science."

Furthermore, in a June 2005 article in the British Catholic magazine *The Tablet*, Fr. Coyne wrote, "If they respect the results of modern science, and indeed the best of modern biblical research, religious believers must move away from the notion of a dictator God or a designer God." Instead, "God in his infinite freedom continuously creates a world that reflects that freedom at all levels of the evolutionary process to greater and greater complexity . . . He is not continually intervening, but rather allows, participates, loves."

The text of a lecture Fr. Coyne delivered at Palm Beach Atlantic University in West Palm Beach, Florida, on January 31, 2006, was published in *Catholic Online*. In this text, Coyne reiterated the points previously reported in the press and added the observation that the intelligent design movement "actually belittles God, [and] makes her/him too small and paltry." Further, Fr. Coyne described Cardinal Schönborn's editorial as "tragic" and "in error."

Perhaps the clearest and most extended statement of present-day Roman Catholic doctrine about evolution comes from Father Rafael Pascual in an interview published by the Vatican's Zenit News Agency in December 2005 and featured on Cardinal Schönborn's Web site. Father Pascual states, "Evolution, understood as a scientific theory, based on empirical data, seems to be quite well affirmed, although it is not altogether true that there is no longer anything to add or complete." He goes on to say that "the Bible does not have a scientific end but rather a religious end. Therefore, it would not be correct to draw consequences that might implicate science, or respect for the doctrine of the origin of the universe, or about the biological origin of man . . . for the Church, in principle, there is no incompatibility between the truth of creation and the scientific theory of evolution. God could have created a world in evolution, which in itself does not take anything away from divine causality."

Concerning humans specifically, Fr. Pascual writes, "on the question of the origin of the human being, an evolutionary process could be admitted in regard to his corporeal nature, but in the case of the soul, because it is spiritual, a direct creative action is required on the part of God, given that what is spiritual cannot be initiated by something that is not spiritual." Fr. Pascual then observes, "On the kind of relationship that the Church promotes with the world of science, John Paul II said the collaboration between religion and science becomes a gain for one another, without violating in any way the respective autonomies." I think Fr. Pascual has expressed the situation perfectly.

The Catholic News Service reports that even an evolutionary biologist has been published in the Vatican newspaper *L'Osservatore Romano*. The article, written in 2006 by Fiorenzo Facchini, a professor of evolutionary biology at the University of Bologna in Italy, states, "If the model proposed by Darwin is held to be inadequate, one should look for another model. But it is not correct methodology to stray from the field of science pretending to do science." The article said that a Pennsylvania judge had "acted properly" when he ruled in December 2005 that intelligent design could not be taught as science in schools. "Intelligent design does not belong to science and there is no justification for the pretext that it be taught as a scientific theory alongside the Darwinian explanation." According to Facchini, "Catholic teaching says God created all things from nothing, but doesn't say how." That leaves open the possibilities of evolutionary mechanisms like random mutation and natural breeding. The article continues, "God's project of creation can be carried out through secondary causes in the natural course of events, without having to think of miraculous interventions that point in this or that direction . . . What the church does insist upon is that the emergence of the human supposes a willful act of God, and that man cannot be seen as only the product of evolutionary processes." The article said, "The spiritual element of man is not something that could have developed from natural selection but required an 'ontological leap.'" Again, I agree completely. I feel a sense of relief, even joy, to see this convergence of views.

In these contentious times, no one writing on behalf of Roman Catholic doctrine, even Cardinal Schönborn, is contesting the basic facts of evolution—that all of life is united in one family tree and that species change over generations of time.

In the homily to start his ministry, Pope Benedict XVI said, "We are not some casual and meaningless product of evolution. Each of us is the result of a thought of God. Each of us is willed, each of us is loved, each of us is necessary." As products of evolution we are not casual or meaningless. I deeply thank God for willing me and for loving me. I thank him too for guiding me into evolutionary biology.

9

To-Do List for Theorists

O HAVE A FAIR AND BALANCED VIEW of evolutionary biology today, we need to see not only its successes but also its current limits. This chapter focuses on some of those limitations, specifically on difficulties that arise because the concept of an "individual" is not as simple as it seems.

This chapter is intended to do more than merely provide balance. Correcting an overemphasis on the individual will redirect the philosophical image of Darwinism away from celebrating the selfish and toward concern for the social. I think this new direction is inevitable as evolutionary biology increasingly grapples with species and life-forms that defy easy descriptions of what an individual is. Coincidentally, this new direction is likely to lead to a science more friendly to

Christian goals of social justice than are present-day social interpretations of Darwinism.

In this book I've been assuming that you, my readers, are broadly representative of Americans in general. Now I wish to speak directly to readers who are evolutionary biologists, whether researchers or teachers. I'd like to encourage you to leave the safety of the topics and species we understand so well and to venture into the species and social systems that are problematic for us. I believe the general public doesn't expect us to have all the answers and that as long as we have some of them we're doing a good job. There's no reason to circle the wagons because of litigation against teaching evolution in a few places in the United States. If we enlist the support of the general public for our quest to understand how evolution happens in species beyond those customarily studied, we'll defuse some suspicion underlying the litigious hostility we face and everyone will benefit. Earlier, I said that evolutionary biologists had correctly figured out only 90 percent of what's going on. What's missing? Well, let's start with what's correct and work out from there.

Is ordinary evolution easy to see? Yes, but you have to travel. If you stay in one spot, the plants and animals around you are not likely to change. You may see environmental change—a fire or flood may allow different birds to come in, or bring special bugs to chew the new leaves—but the birds and bugs themselves are not likely to change much in size or color before your eyes, which is what you would have to see to witness evolutionary change in real time. That said, if you're using herbicide to control weeds or poison to control ants, you may see the weeds in your backyard or the ants in your cupboard evolve tolerance to the chemicals you're using. This happens because only the weeds and ants immune to the chemicals survive to breed and thus slowly take over your backyard or house. For the most part, to see evolutionary change solely at a single location, you'd have to wait a long time, longer than most people have.

So, is evolution hard to see with your own eyes? No, again, not if you travel. You need only compare the way the plants and animals you're used to at one spot, say, your backyard or the street, look and behave with the way they look and behave somewhere else. This is what clued Darwin in to evolution in the first place, comparing the finches in South America to those in the Galápagos and seeing slight but real differences in the shape of their beaks, their body size, and the color of feathers.

Evolution textbooks are filled with the authors' favorite examples of evolution. Some famous moths from England make it into all the books, including one I wrote long ago. The moths started out speckled, to blend in with lichen-covered tree trunks. Then came the Industrial Revolution, dumping black soot onto the trees and making these moths stand out as though shouting "eat me" to the birds. Well, eat them the birds did. Then mutations turned up that produced black pigment, and the black moths survived better to breed, and after some generations the moths became black. Later, when the factories had to clean up their smoke, the trees were again covered with lichens, and the moths evolved back to their earlier color. All the details of this case have been worked out. Presenting this one case over and over again in textbooks gives the impression that although evolution through natural breeding may happen, it is rare and restricted to just a few cases. Yet when you travel around, you see evolution all over.

When dogs and cats are released into the countryside, they go native, or feral, after several generations. Just try petting a feral dog or cat, giving it some table scraps. You'll be lucky to emerge unbitten or unscratched. And look how long their ears have gotten and how fast they now run. They've evolved to be wild animals again and can breed under natural conditions, whereas those that wag their tails and purr sweetly have the best chance to breed when living around people. We have also learned that DDT causes flies and mosquitoes to evolve resistance. Antibiotics cause bacteria to evolve drug resistance. The list goes on and on.

All biologists have their own favorite cases of evolution in the plants or animals they work with. Mine is the lizards that live throughout the island of Dominica in the West Indies. Some parts of this island, or its rocky foundation, have been around for tens of millions of years. Think of the island as an oval about fifteen miles long and five miles across, with mountains in the center. The lizards change in size, shape, and color on a mile-by-mile scale over the whole island. Drive around the island, stopping every mile to get out and look for lizards by the roadside or in nearby woods, and you'll find that they gradually change with every mile, from gray to brown to green, with white spots to black spots to white rings to mossy weblike designs.

My hunch about why these differences have evolved is that they represent "dialects" of "body English." The white-crowned sparrows around San Francisco Bay have evolved different song dialects. Lizards don't sing, of course; they rely on body color to communicate, in much the same way sailors used flags to communicate before radio was invented. I theorize that for a lizard to have a dialect it needs a body marking special to its location. The lizards around Dominica may be speaking many dialects of body English, just as people and even birds develop dialects after inhabiting a region for a long time.

You can also check out any young island in the West Indies like St. Kitts produced by a volcano around two million years ago. All the lizards on that island are the same, presumably all resembling the first lizard colonists who arrived when the island first emerged from under the ocean. The differentiation into the local breeds seen on Dominica has taken a long time—much longer than two million years.

If you're not as excited about lizards as I am, pick your own favorite plant or animal, and travel around. You'll see your species change from place to place in accord with how long it's been there and what your organism particularly needs to do at each spot in order to breed there successfully. If these examples whet your appetite for more, many excellent documentaries about evolutionary change are available on PBS stations, the Discovery Channel, and the BBS; from the National Geographic Society; and in exhibits at museums in all major cities. Beautiful books, including some from Time-Life, are available on evolution, too.

I go into all this because I hope you can appreciate how hard it is for people like me who love plants and animals to contain our enthusiasm for sharing with you what we know about our organisms. Unfortunately, it's all too easy for us to stop here and not go on to talk about the situations that don't automatically fit the evolution by natural breeding narrative.

The situations that don't fit easily into present evolutionary theory all pertain somehow to the idea of an "individual." The problems are first, do individuals really work as individuals or as part of a larger whole, and second, what is an individual anyway? These problems have confusingly similar names: "individualism" and "individuality."

Individualism is the premise that we can credit an animal's success at breeding to what it does by itself—it deserves all the credit for its own success. The assumption in evolutionary theory is that an animal reaps the evolutionary reward of its own labors. But when an animal's success relies on others, evolutionists don't know how to apportion credit to everyone involved, and evolutionary theory grinds to a halt while people figure out what to do about this.

Individuality is a somewhat different problem—it's when we can't figure out what an individual is to begin with. When we're thinking about mammals—our pets, livestock, and even ourselves, what counts as an individual cat, cow, or person is obvious. But as we move out into the rest of living creation, the distinction gets less and less clear. What about grapevines made from branches planted in the ground that grow into plants of their own? Are all the zinfandel grapevines of the world one giant individual with many pieces, or are they many individuals that are clones of each other? The answer depends on whether you define an individual as everything coming from one seed or as something confined to a distinct bodily package. Thinking about this problem has been known to give biologists migraines.

Let's expand on these problems to give you a sense of where evolutionary research may go in the future and how evolutionary theory is still a work in progress.

Big problem #1: Individualism. Are the breeding units groups or individuals, or something in between? For moths, success or failure at breeding seems to depend only on individual performance. If an individual moth finds more food and evades more birds than other moths do, then it breeds more than the other moths and its genes increase in the next generation relative to the others. Simple enough.

Consider ants or bees instead. In an ant colony, only a few ants—the queens—do all the breeding. The worker-ants are their daughters and don't breed themselves, but help their mothers breed. In this case, successful breeding refers to the total number of offspring the colony produces, not to the offspring that an individual worker produces, which is zero. Suppose an anteater wanders by, sniffing around to discover an ant nest. Gobbling up a few individual worker-ants doesn't hurt the breeding success of the colony. But if an anteater gobbled up the queens, it would wipe out the breeding success of the entire colony.

Many species, perhaps even most, fall between the extremes where the individual is everything, like moths, to where the individual is nothing, like ants. Instead, breeding in most species relies on a biological infrastructure provided by the animal's social system.

Wolves forage in packs, fish swim in schools, birds fly in flocks. The individuals do their breeding within these social systems. Only by participating in these social infrastructures can individuals obtain food to raise offspring and safety from discovery by predators.

An animal works sometimes as an individual and sometimes as a member of a group. Sometimes breeding success reflects group ability more than individual ability. No one has figured out how to think about natural breeding at these multiple levels extending from the individual through the social group up to the species. The classic Darwinian narrative emphasizes individualism only and doesn't appreciate the multiple levels on which overall breeding success depends.

This problem, called the "levels of selection" problem in biology, points to the poor job biologists are doing with understanding cooperation in nature. Yes, lots of competition occurs in nature, but competition is almost always mixed in with cooperation, too. Cooperation is especially prominent at levels above the individual. An animal's social infrastructure generally includes ingredients of cooperation as well as competition. Because animals are invariably embedded in a social infrastructure, the libertarian ideal of individualism is rarely biologically realistic. Moreover, nature never says, "Every man for himself." Humans say that, hoping nature will excuse their selfishness.

Big problem #2: Individuality. The second big

problem of evolutionary biology today is dealing with the breakdown of how an "individual" is defined. A biological individual is understood to be an organism that consists of one body with one genome. A genome is an entire package of genes transmitted together from generation to generation. The problem of individuality arises when one genome produces multiple bodies or when multiple genomes coexist in one body.

In many species of both plants and animals, one genome produces many bodies. A poplar tree sprouts many trunks from its roots. A grove of poplar trees is usually one genetic individual in the sense that every trunk has sprouted from the same seed, but many bodies in the sense that each trunk lives a life of its own. In the ocean, many animals bud off duplicates of themselves that then live as colonies whose members are physically attached to one another—for example, corals, tunicates, hydra, bryozoans, and sea anemones. The Portuguese man-of-war jellyfish is a floating colony, like a space station of animals drifting through an oceanic universe.

In plants and in many animals, numerous species have an alternation of generations in which an asexually reproducing phase leads to a sexually reproducing phase, followed again by an asexual phase. The individual created at the sexual stage by the union of an egg and a sperm then creates many bodies, each of which contains its genome. These asexually created bodies

can then mate to produce offspring sexually. This phenomenon of multiple bodies arising from one genome
is extremely common, and its implications for evolution are poorly understood.

Conversely, many genomes can coexist in a single
body. This is sometimes called endosymbiosis, when
one cell lives inside another. Great examples are the
corals. Corals are colorless animals that allow algae to
enter their bodies. The algae give the corals their color.
The algae photosynthesize and supply nutrients to the
coral in return for living in its tissues. The algae can
come and go. When the algae leave, the corals then become transparent again, a process called coral bleach-
ing. Many different strains of algae can enter the corals.
When the algae are inside the corals, the coral individuals contain two genomes, their own and the algae's.
From the standpoint of reproduction, however, the
coral and algae are a unit, joined at the hip, so to speak,
because the algae live inside the cell membranes of the
coral cells.

Consider the totality of genes in a coral cell, both its
own genes located in its nucleus plus the extra genes in
the algal cells. This total combined genome is what the
coral cell has to work with. Next, ask what is the source
of genetic variation for a coral cell. Well, there's the
usual random mutation within its nucleus and within
the algal cell's nucleus, too. In addition, a coral cell can
discharge the algal cells and allow others in instead.

To-Do List for Theorists

There can be a biochemical negotiation between coral and algae resulting in the coral acquiring the genes it needs in return for providing what the algae require for their habitat. This is not random mutation. Instead, this process allows a coral cell to steer its own genetic variation in a direction that leads to the most breeding success under present conditions, a sort of directed mutation. This is a fundamental departure from the neo-Darwinian narrative of natural breeding building solely on randomly generated genetic variation.

Who cares about corals? Are there any other examples? Yes, us. We've been told in grade school that each of us has half of our genes from our mother and the other half from our father. Well, that's not quite true. In each of our cells, most of the genes are in the nucleus, and other genes are in tiny football-shaped structures called mitochondria. The biologist Lynn Margulis discovered that the mitochondria were once free-living bacteria. Therefore, each of our cells contains two distinct genomes—the genes in the nucleus that are inherited equally from our mother and father, plus the genes in the mitochondria, which are all inherited from our mother. Overall, more of our genes come from our mother than from our father. Many biochemical functions, and even individual proteins, depend on products from both nuclear and mitochondrial genes, so these two classes of genes must work well together.

This raises the question of whether we're really the

individuals we think we are. We, ourselves, are basically colonies. As everyone knows, many bacteria live in the gut, trachea, and mouth—these bacteria are in the body but not actually within the cells. We do need to cooperate with our intestinal flora, for otherwise we suffer indigestion or diarrhea. But the cooperation needed between our nuclear and mitochondrial genes is much more intimate, because both sets of genes must work together within each of our cells or else we die. We have no idea of how our ancestor cells early in life's history interacted with the available bacteria at the time to recruit them to join in a common mission. Whatever happened, it wasn't simply the acquisition of new genes through random mutation. It must have involved some cooperative assimilation of an entire packet of genes bundled up in the genome of the ancestral mitochondria.

Rethinking the place of individualism and the criteria for individuality in nature will produce an evolutionary biology having a different philosophical flavor from present-day neo-Darwinism. This will be an evolutionary biology of interdependency, a narrative that we are all of one body and that furthers a vision of Christian community within nature.

This chapter has presented what I think are genuine problems with evolutionary biology today. We now move to a discussion of "intelligent design," a political movement that claims to have identified other problems with evolutionary biology. Intelligent design makes up problems with evolutionary biology that don't exist, while ignoring the deep problems that do exist, like those we have just explored, those concerning what an "individual" is.

10

Intelligent Design

N 1991, PHILLIP JOHNSON, A PRO-
fessor at the University of California–
Berkeley Law School, published *Darwin on Trial,* a short book advertising
that he would look "at the evidence for
Darwinistic evolution the way a lawyer would." I reviewed the book in 1992 at the request of the journal
Ecological Economics and concluded that the book added
"little light to the creationist/evolutionist debate, but
its sarcasm and condescension do add heat." Nonetheless, I predicted that "students of the environment will
have to be somewhat conversant with it for the near
future."

Little did I know the book would kick off the current organized and highly funded political challenge to
teaching evolution in public schools. This challenge is

now known as "intelligent design"—it is the search for evidence of God through science. God's fingerprints, so to speak, consist of instances of intelligently designed features in organisms, which, if found, would constitute scientific evidence that God exists. The quotations from President Bush and Senator Frist mentioned at the start of this book trace to the present-day political success of adherents to the intelligent design movement.

A tone of indignation emboldens the intelligent design movement as though evolutionary biologists were foisting anti-religious dogma on innocent children. Johnson writes, "Make no mistake about it. In the Darwinist view, which is the official view of mainstream science, God has nothing to do with evolution." This is simply not true. I cannot say it too many times: Scientists as a group, including evolutionary biologists, do not take any position one way or the other on the role of God in evolution, and a great many scientists, myself included, are active Christians and proud of being both a Christian and a scientist at the same time.

Is intelligent design the same as creationism? At least in principle, it is different. Johnson himself writes, "I assume that the creation-scientists are biased by their pre-commitment to Biblical fundamentalism." He continues, "I am not interested in any claims that are based on a literal reading of the Bible, nor do I understand the concept of creation as narrowly as Duane Gish does." (Duane Gish is a leading spokesman for the creationist movement.)

Theoretically, the difference between creationism and intelligent design is this: With creationism, you start out believing in God and in a particular reading of Genesis, and then stay a believer as scientific data supposedly confirm your position. With intelligent design, you start out as agnostic and then later become a believer because the scientific data supposedly convinces you that an "intelligent designer"—that is, God—exists.

The legal status of presenting creationism as science is now settled law—it's illegal. Here's the basic principle. It's against the law to preach religion in public schools—not just any law, but the U.S. Constitution, the highest law of the land. Preaching religion is perfectly legal in private schools, in places of worship, in homes, and on the street corner, but not in government facilities. According to the Supreme Court in 1987 (*Edwards v. Aguillard*), presenting creationism as science is preaching because it amounts to saying that a specific religious belief, namely, the Christian God plus a particular reading of Genesis, is true. Therefore, presenting creationism as science in public schools is breaking the law because it is preaching.

Teaching intelligent design might not be against the law because, hypothetically at least, you could be completely neutral about the existence and nature of God and let the data do the talking. That wouldn't be preaching; it would be teaching and might be legal. If so, whether intelligent design should be included in any science class depends on whether there's anything to it, whether any real science backs up claims from the intelligent design movement.

Most scientists would go further and say it's absolutely impossible to prove the existence of God with scientific data—that God belongs to a spiritual realm and scientific data to a material realm. If so, intelligent design is destined to fail as a scientific project because no conceivable data could ever prove the existence of God. Intelligent design would then be incorrect, but not illegal. What's illegal is preaching a religion in the public schools, whether in a science course, a philosophy course, or anywhere in the curriculum. In contrast, teaching bad science is merely incompetent.

But any conceptual distinction between creationism as religion and intelligent design as science has now been hopelessly compromised. Although the intelligent design movement initially disassociated itself from fundamentalist Christians, accusing them of bias, fundamentalists today are the principal supporters of intelligent design. Many fundamentalists have taken the phrase "intelligent design" as a synonym for creationism without knowing what it is. In a recent court case concerning the school board of Dover, a small town in Pennsylvania, the judge observed that school board members who favored teaching intelligent design didn't know what it was and one couldn't even pronounce it correctly. The school board had originally wanted to teach creationism in science classes and hoped to get around the law by re-labeling creationism as intelligent design. The judge said, no, creationism by any name is still illegal.

Creationism has become inextricably commingled with intelligent design. The intelligent design advocates brought this situation upon themselves. One of these advocates, Paul A. Nelson, wrote in the Vol. 24, No. 4, 2002 issue of the *Christian Research Journal,* "God could have created everything in six 24-hour days—or not. The fundamental point is to allow for the possibility of design. The scientific narrative of design—when God acted, and how—might capture any number of competing theories." This "big tent" approach of the intelligent design movement invited a wide political coalition to join in opposition to teaching Darwin in the schools. The big tent opened the way for creationists to sign on to intelligent design, even though they didn't know what it was. They assumed that any group opposed to teaching evolution in the schools was automatically an ally—the enemy of my enemy is my friend.

Intelligent design advocates are now trying to distance themselves again from creationism. In January 2006, a near repeat of the Pennsylvania case was about to unfold in a rural community in the southern Sierra Nevada of California. The El Tejon School Board had approved a course that was originally a creationism course but was renamed to be a course on intelligent design. A lawyer for the Discovery Institute, a think tank in Seattle that promotes intelligent design, wrote to the school board saying that the course should be pulled because it was primarily a creationism course, not an intelligent design course. The course was then discontinued. This shows the big-tent strategy is now being abandoned.

It's probably too late to disentangle intelligent design from its commingling with religion. For example, intelligent design advocates continue to cite Cardinal Schönborn as an ally. Yet, as we saw earlier, the cardinal's position that design belongs in science courses is not about conducting science but about introducing a particular religious philosophy into science courses. This too would be found illegal if it were challenged in court, because it would be preaching, not teaching. To reestablish their credibility as offering a scientific alternative to Darwinian evolution, intelligent design advocates need to sever all the political connections they have built through their big-tent strategy because their other tent mates want to use science classes as a forum for preaching, not teaching.

Given that intelligent design is on the ropes for now, many of its opponents look to a knockout punch that will silence its advocates once and for all. I don't think that will happen, regardless of further court victories and litigation. For any chance of peace, I believe we need to consider both the scientific and the religious content of intelligent design and see why it is a non-starter as science and, more important, why it is not helpful to Christianity.

Let's turn to the science behind intelligent design. The intelligent design movement concedes much of evolutionary theory to begin with. Johnson writes, "There is no reason to doubt that peculiar circumstances can sometimes favor drug-resistant bacteria, or large birds as opposed to small ones, or dark-colored moths as opposed to light-colored ones." Changes in size, shape, and color are called microevolution, and he goes on to say "everyone agrees that microevolution occurs." So what's the beef?

The basic point to intelligent design is that animals possess some structures that God made from scratch and just put there. How do we find these God-given structures that arise "abruptly," according to the word that intelligent design proponents use? They are structures that are "irreducibly complex," which means they need all their parts to work. If any part is taken away, the whole structure fails. The evolution of such structures would seem to be unattainable through a gradual bit-by-bit modification of preexisting structures, as envisioned by the neo-Darwinian narrative of natural breeding building on independent genetic mutations. Therefore, intelligent design advocates think God—the intelligent designer—must have just put them there, already made.

Johnson and later writers in the intelligent design movement such as Michael Behe, a biochemistry professor at Lehigh University, have mentioned a potpourri of examples including the eye, the respiratory system, the feather, the proteins that cause blood clotting, and the bacterial flagellum. Although we all know what an eye, a lung, a feather, and blood clotting are, a bacterial flagellum may seem rather obscure. Some bacteria have a thick hair sticking out their back end that they wiggle to move forward. It works somewhat like a corkscrew, and the propulsion reminds one of a riverboat or gondola with a paddle at the stern. This flagellum is supposed to be irreducibly complex.

So where did animals get their eyes, or bacteria their flagella? The intelligent designer gives them these structures out of the blue. Intelligent design advocates refuse to specify any scenario for when and what occurred, so we are left to make them up ourselves, to imagine animals stumbling around blindly until the intelligent designer plops eyes into empty eye sockets. Or upon seeing bacteria stuck in the mud, the intelligent designer seeds the mud with unattached flagella. The flagella wiggle around and bump into some bacteria, whereupon they attach to one another and live together forevermore. The unwillingness to specify a hypothesis for what has supposedly happened is a source of frustration even to scientists favorably disposed to intelligent design because one can't begin to test a nonexistent hypothesis.

The intelligent designer makes miracles. The many irreducibly complex structures that organisms are said to possess constitute miracles their lineage has supposedly accumulated through time. This makes for a lovely creation myth. Unfortunately, there's not a shred of evidence for this myth.

Instead of finding evidence for their own theories, intelligent design advocates spend their time attacking neo-Darwinism. This is a basic error in logic. Neo-Darwinian evolution versus intelligent design is not an either/or proposition. As a matter of pure logic, Darwin and intelligent design could both be wrong, Darwin right and intelligent design wrong, Darwin wrong and intelligent design right, or both right. Saying that neo-Darwinian evolution and intelligent design are mutually exclusive is setting up what many call a "false duality."

Scientifically, the hypothesis of intelligent design will have to stand or fall on its own merits. Intelligent design advocates are wasting time by attacking the neo-Darwinian account of how evolution happens, because some third possibility could always be proposed, or a fourth, or fifth . . . Neo-Darwinists may modify their hypotheses in light of new data, say, by considering mechanisms of inheritance unknown to Mendel; and intelligent design proponents could also change their positions, say, by abandoning the flagellum and seizing on some other structure as a candidate for something supposedly too complex for neo-Darwinian evolution to explain.

Furthermore, neo-Darwinism *can* account for complex structures. When you get together eye experts, lung experts, feather experts, flagellum experts, blood clotting experts, and so on, it always turns out that they can suggest plausible neo-Darwinian scenarios for how these structures originated. The details will surely put you to sleep—the bacterial flagellum may have evolved from a biochemical subsystem called a Type III secretory system.* The various proteins involved in blood clotting are not irreducibly complex, because parts can be removed and clotting still occurs. One of the proteins has been omitted in the blood of whales and

---

*A Type III secretory system is a ring of proteins in the bacterium's cell membrane surrounding a pore though which a bacterial cell extrudes poisonous proteins into the cell of the host it's infecting. A bacterial flagellum has a ring of proteins at its base. Some genes for the bacterial flagellum are closely related to those for the Type III system. A hypothesis is that proteins called flagellin that comprise the flagellum tail are extruded through the pore and solidify to become the tail. This illustrates how a preexisting structure with a known function, secretion, might acquire a new function, locomotion, by the addition of small changes.

dolphins, but their blood still clots, and three parts are missing from the blood of puffer fish, and their blood clots, too. How much do you really want to know about the bacterial flagellum and how blood clots? If you do want to know more about complex structures, a good place to begin is an article by Sean Carroll in the November 2005 issue of *Natural History Magazine,* published by the American Museum of Natural History in New York. Still, you may wonder how discussing the technical details behind bacterial flagella and blood clotting can possibly offer a route to discovering God. I can't help feeling that this approach misses the whole point to Christian faith.

I know people may be uncomfortable with how dismissive evolutionary biologists are when discussing intelligent design. This discomfort alone is enough to turn some against evolutionary biology, regardless of the intellectual points being made. Is this dismissiveness just attitude, or is there some basis to it?

Intelligent design strikes evolutionary biologists as naïve. Let me try to give you a sense of why biologists view intelligent design as a waste of time. When I was a twenty-two-year-old graduate student at Harvard, I assisted in a course on invertebrate zoology—that's all about animals without backbones. The professor was the world's expert on echinoderms, starfish and the like. One day in lecture he announced that he had seen a coral-reef fish whose green color was so beautiful it couldn't possibly be explained by natural selection. My jaw dropped, and I asked him later if he meant it. He said, yes, such an emerald green was beyond what natural selection could do.

Well, biologists generally know that an iridescent green pigment is easily produced through ordinary evolutionary processes, and many species on land and sea have evolved a beautiful green color. But this green-fish case was to be one of many instances when I would hear an experienced biologist, or myself, marvel at some amazing trait in the animal kingdom and wonder how ordinary evolution could have produced it.

My favorite marvel is a chameleon's tongue, several body lengths long, which sticks out in a flash to catch bugs far away. How could this possibly evolve? A medium-sized tongue would be useless, because a lizard couldn't get close enough before the bug flew away. So, how does a chameleon's tongue get to be so long? Offhand, I don't know, but appeal to an "intelligent designer" is simply a cop-out, a way to avoid doing the hard work of figuring out whether there are any relatives with medium-sized tongues, what else tongues are used for, what fossil tongue lengths are, and so forth.

There's nothing new about the idea of a trait being very complex. We biologists hear it all the time. Biologists who have studied the history of their subject will have encountered it there, too—"intelligent

design" was framed by St. Thomas Aquinas in the thir-
teenth century and notably advanced by the Reverend
William Paley in his influential 1803 book *Natural
Theology*. Darwin even wrote of his admiration for Paley,
revealing a cordiality in the face of disagreement that is
all too rare today. So, when the intelligent design folks
announce with great fanfare that the bacterial flagellum is too complex to be explained by natural selection
acting on random mutation . . . well, it's hard for evolutionary biologists to suppress yawns. Yes, the bacterial
flagellum is beautifully complex. But a lot of nature is
marvelous. The whole of nature is marvelous, as befits
God's creation, not just little bits and pieces here and
there.

Furthermore, I've learned not to underestimate the
power of natural breeding acting on random mutations
to yield very complex structures. I once did a study
with computer scientists to develop a computer program that a lizard might follow when making decisions
about what prey to eat and what to ignore. This could
be the computer program underlying the minute-by-
minute behavior of a lizard outdoors in nature.

We tried to develop the lizard-feeding program by
mimicking how evolution would do it. We started out
with a population of randomly generated computer
programs for feeding—some of these described really
clumsy animals. Then we chose the best 10 percent of
them and let them "breed" by mixing them together

and reassembling the pieces into a new generation of computer programs. Then we took the best and repeated the process for about ten generations. By then we had evolved very successful computer programs from randomly assembled pieces. The lizards, who all started out more clumsy than a baby eating pureed carrots, had evolved into connoisseurs of fine dining whose technique Emily Post would have envied.

But what I found most interesting was looking at how these evolved computer programs worked. We imagined that the computer programs assembled through this evolutionary process would be models of fine organization. To humans, good computer programs have a certain structure, much like a good recipe in a cookbook, with a list of ingredients, followed by instructions for preparing the main parts, ending with the final assembly. But the programs for our very smart lizards were an unorganized jumble of instructions that nonetheless worked together perfectly. And that's just how our own genome is organized. Our genes are not laid out across our chromosomes in a neat order. The genes for our ears, eyes, noses, and throats are scattered across our twenty-four chromosomes in a mishmash, haphazard order. This is the signature of complexity emerging from natural breeding acting on random mutations. This is life.

Each of us in evolutionary biology has learned, often to our amazement, that extraordinary complexity and beauty do emerge from natural breeding acting on random mutations. The guidance that leads to this astonishing complexity is the sustained direction provided by the natural husbandry component of the evolutionary process. One cannot underestimate this. So, when a Johnny-come-lately political movement like intelligent design turns up from folks without much experience in evolutionary biology and without any scientific data to support their theory, it's hard to take seriously.

What would intelligent design proponents need to do to make their program scientifically credible? I would like to see four scientific points addressed. Intelligent design scientists need to publish an objective procedure to screen for complexity so that the five best-case candidates for irreducibly complex traits can be defined for analysis. Then, they need to explicitly state and present direct evidence for specific hypotheses about when the traits first appeared and in what form. Next, they need to demonstrate that natural breeding acting on random mutations does not account for these best-case complexity candidates. Finally, should existing evolutionary theory prove inadequate, then intelligent design scientists need to show that no possible material modification of the theory can be made that would account for the candidate traits. If all four criteria are met, then I would say that the intelligent design program has succeeded scientifically. Until then, it's hot air.

Intelligent Design

I'm completely in favor of "teaching the controversy," as advocates of intelligent design often demand in the name of "fairness," but the controversy has to be real, not made up. I believe our biology courses should teach about the limitations of evolutionary theory to deal with species in which an "individual" is not well defined, with species in which the individuals work in teams and social groups rather than as solitary agents, and with species, as I'll cover in the next chapter, in which the sex and gender roles are different from those Darwin assumed were standard. These issues are real problems for evolutionary theory today, not the make-believe problems concocted by the intelligent design folks.

For most scientists, the matter of intelligent design is already completely settled—intelligent design has no data going for it, and intelligent design advocates are wrong in charging that complex structures can't be explained through neo-Darwinian narratives. For me, though, I'm more concerned that the religious assumptions behind intelligent design are not receiving attention. Intelligent design asks you to believe in God on the basis of miracles.

What do you think of miracles? I know that if I were into miracles, I wouldn't mess with the bacterial flagellum. I'd go for the big ones: Jesus curing lepers, the deaf, the blind, and the dumb from Matthew, Mark, and Luke, or making wine from water, or raising

Lazarus from the dead in John. My favorite is Matthew 15:36–38, in which Jesus feeds thousands of people from seven loaves and still has seven loaves left over: "And he took the seven loaves and the fishes, and gave thanks, and brake them, and gave to his disciples, and the disciples to the multitude. And they did all eat, and were filled: and they took up of the broken meat that was left seven baskets full. And they that did eat were four thousand men, beside women and children." Now that's a big miracle. Forget about the bacterial flagellum. How do you feel about feeding thousands of people from seven loaves of bread?

I don't know if the seven loaves event happened as described. I've never seen anything like it in my experience. On the other hand, I'm not prepared to say that St. Matthew, or St. Mark who describes the event in Mark 8:1–9 and St. John who describes it in John 6:5–13, are mistaken. Nor are the multitude of others who presumably could have contradicted the account. But then I wonder, why do I need to take a position? Does my faith in God and in Jesus' teachings have anything to do with miracles? No. Jesus' teachings about generosity, kindness, love, and inclusion of all don't depend one whit on miracles. Jesus himself does not require that we believe in miracles. In fact he urges us *not* to rely on miracles as the path to faith.

Mark 8:11–12 records how Jesus was asked to prove himself by displaying some sign from God, some miracle: "And the Pharisees came forth, and began to question with him, seeking of him a sign from heaven, tempting him. And he sighed deeply in his spirit, and saith, Why doth this generation seek after a sign? Verily I say unto you, There shall no sign be given unto this generation." Jesus carried out miracles not to prove that God exists or that he was the son of God, but to accomplish some genuine purpose—feeding people who were hungry or curing people who were sick. In John 6:26, Jesus instructs his followers not to focus on miracles: "Verily, verily, I say unto you, Ye seek me, not because ye saw the miracles, but because ye did eat of the loaves, and were filled."

Another aspect to miracles is that they are unconvincing in themselves. In John 6:36, Jesus acknowledges those who still don't believe even after having seen him miraculously feed the multitude. He goes on to say in John 6:40 that those who nonetheless do believe will be raised up to heaven: "But I said unto you, That ye also have seen me, and believe not . . . And this is the will of him that sent me, that every one which seeth the Son, and believeth in him, may have everlasting life: and I will raise him up at the last day."

Even after his death, Jesus continued to downplay miracles. After his resurrection, Jesus appeared to a group of his disciples that didn't happen to include Thomas. When the disciples who did see Jesus told Thomas of Jesus' resurrection, he doubted and said he wished to put his fingers in Jesus' wounds from the cross before he would believe. Thomas said, "Except I shall see in his hands the print of the nails, and put my finger into the print of the nails, and thrust my hand into his side, I will not believe. And after eight days . . . came Jesus . . . Then saith he to Thomas, Reach hither thy finger, and behold my hands; and reach hither thy hand, and thrust it into my side: and be not faithless, but believing. And Thomas answered and said unto him, My Lord and my God. Jesus saith unto him, Thomas, because thou hast seen me, thou hast believed: blessed are they that have not seen, and yet have believed." (John 20:25–29)

Miracles are at best a crutch for those, like Thomas, of little faith. Miracles by themselves don't convince a skeptic to become a believer, and their absence shouldn't dissuade a true believer. In the terminology of logic, miracles are neither necessary nor sufficient for belief in God.

Focusing on tiny miracles like the bacterial flagellum winds up diverting us from the miracles in scripture. Intelligent design offers a backdoor route to belief in God. It turns you away from the Bible to look at scientific data. If you don't believe the Bible's miracles, check out the bacterial flagellum. Instead, all of creation is the miracle, not bits and pieces here and there.

Finally, I suggest the phrase "intelligent design" is pretentious. Who are we to give God an intelligence

test, to measure his IQ? Saying that God is "intelligent" invites the sin of idolatry.

I ask that conservative Christians think about what intelligent design really is and question whether it's worthy of their support. If intelligent design succeeds as a movement, Christianity will be hurt. Intelligent design says the facts of nature offer a better testimonial to God than the Bible does. It will substitute science for the Gospels. The weekly sermon will be about new data on the biochemistry of flagella, not on Jesus' parables.

Let's conclude with whether intelligent design should be taught in our schools and, if so, where. Intelligent design doesn't qualify as adequate science. "No data? Not science," pretty much says it all. Intelligent design makes up problems with evolutionary biology that aren't there, like explaining eyes, ears, and flagella, while ignoring problems that do exist, like individualism and individuality discussed earlier, and gender and sexuality, which I'll discuss next.

Nonetheless, what I've learned from the political success of the intelligent design movement is the need to teach what makes bad science bad. I favor teaching intelligent design as an example of bad science. I don't favor simply tossing both Darwin and intelligent design to students and saying, "Hey, you decide." That would be irresponsible. We have to say why intelligent design is junk science. The decision by Judge John E.

Jones III in December 2005 concerning the Dover, Pennsylvania, school board might be assigned as a reading, specifically section E4 (pp. 64–89), where he ruled that the intelligent design curriculum presented in this case was not science.

I know we teachers begrudge giving up precious class time to bad science when it means we have to omit some good science. Class time is the only chance to tell our children about electrons and why there's electricity, about radio waves and why there's television, about the binary code and why there are computers, about continental drift and why volcanoes and earthquakes happen, about the Bernoulli principle and why airplanes fly, about chemical bonds and why gasoline powers cars, about carbon structure and why sugar dissolves in water and fat doesn't, about genes and why dogs don't give birth to cats, about the universal tree of life and why all living things are related, about species changing and why we don't look like our caveman ancestors.

If we add intelligent design to the science curriculum, what goes away—volcanoes, airplanes? But if we keep doing what we have been, our students will continue winding up uneducated about good science versus junk science. Science is not just an accumulation of facts, it's *how* to discover facts and how to explain them. We must teach that science depends on stating testable hypotheses, then actually doing the test, and then standing aside while the tests are confirmed or refuted by other, independent parties. That process produces facts and explanations, not opinions. Assertions that are not testable, or never will be tested, or will be tested by only one group with a vested interest cannot be considered good science. Scientists certainly have opinions, of course. There's a lot of chatter on the way to the end result. Nonetheless, the end results are hard facts of nature and explanations of nature that grab you like the shock of touching a live wire.

I also favor teaching intelligent design in a religious studies curriculum as an example of junk religion. That's the point. When a crossover movement turns up that mixes religion and science like intelligent design does, it must be examined carefully in both domains. Its shallow religion matches its empty science. Its danger to Christianity overshadows its emptiness for evolutionary biology.

I now turn to the final major topic where Darwin and religion intersect: marriage, family, sex roles, and sexuality, including homosexuality.

II

Gender and Sexuality

UR QUEST TO DETERMINE THE strengths and limitations of present-day evolutionary biology ends with what is perhaps the most controversial aspect of Darwin's writings. These writings offer what amounts to a theory of universal sex roles for males and females throughout the animal kingdom. These roles are often thought to apply to humans, as though we too are governed by the rules supposedly evident in animals.

The section of Darwin's writings that pertain to sex roles is called the theory of "sexual selection" and is among the last subjects he wrote about. Darwin worried about the evolution of traits like the peacock's tail, called male ornaments, and the theory he developed for such traits has been extended to become an all-encompassing biological theory for gender roles. This, to my knowledge, is the only part of Darwin's work that is so seriously incorrect that it cannot be updated or revised to make it right.

Some people have asked me why I bother to raise the difficulties faced by the sexual selection section of Darwin's writings. Anti-evolutionists do not point these out and appear unaware of them. Isn't raising this topic giving "ammunition" to the anti-evolutionists? My response is twofold.

I imagine an auditor who finds problems with the financial statement of a corporation must feel as I do. Is there merely some small slipup, or do we have a problem that might balloon someday, sending the company into bankruptcy? By fully disclosing the difficulties now, we avoid any later charges of a cover-up.

Furthermore, as an evolutionary biologist I feel responsible for making the subject as accurate as possible. This is why I urge biologists to devote more attention to species in which the concept of an "individual" is problematic, as we discussed earlier. This is also why I think evolutionary biologists need to get the evolutionary theory of gender roles correct. The sexual selection area of evolutionary biology is currently a mishmash of "big" generalizations with thousands of exceptions and special-case work-arounds. I think the topic needs to be rethought from the ground up.

So what is sexual selection theory? Here's what Darwin wrote in 1871 in *The Descent of Man*: "Males of almost all animals have stronger passions than females" and "the female . . . with the rarest of exceptions is less eager than the male . . . she is coy." This is sometimes true, of course. But Darwin says the passionate-male—coy-female description is true almost all the time and that the exceptions are rare enough not to ruin the general pattern.

Darwin went on to talk about what females were looking for in a mate. This is where the peacock's tail comes from—it's a handsome ornament that female peacocks are supposed to want in a male peacock. Females also want their mates to be tough. Darwin wrote that females choose mates who are "vigorous and well-armed." Over generations, this results in the males evolving to be good fighters, "just as man can improve the breed of his game-cocks by the selection of those birds which are victorious in the cock-pit." Darwin continues, "Many female progenitors of the peacock must . . . have . . . by the continued preference of the most beautiful males, rendered the peacock the most splendid of living birds."

These passages have a familiar ring. They've been picked up throughout popular culture and pop psychology. Here's what *Elle* magazine's February 2005 issue has to say in the column Ask E. Jean: "Males fighting for females is the elastic in the jockstrap of evolution, therefore women are hardwired to 'size up' and appreciate male competition." I am not criticizing *Elle* magazine because the quote is probably a fair, though colorful, summary of what the columnist was actually taught in biology class.

What's fundamentally wrong with Darwin's sexual selection theory is that it assumes males and females begin their relationships from a basic conflict of interest. In neo-Darwinian terminology, females want males with "good genes." Males want lots of sex. These objectives inevitably conflict. Perhaps the conflict can be papered over through subsequent threat and negotiation. If so, male–female cooperation follows after conflict—it attempts to ameliorate male–female conflict present at the beginning. Evolutionary biologists today often describe animal behavior in metaphors such as "battle of the sexes." Well, is the male–female relationship necessarily a battle, throughout nature and in humans as well?

In contrast, I have argued in my recent book *Evolution's Rainbow* that the initial evolutionary objective of male and female is identical—to leave descendants. Males and females begin not with a conflict of interest, but with a shared interest. From the start, they are engaged in a joint venture, to raise offspring as a common investment holding genes from both parents. Sexual reproduction begins as a cooperative venture, and if conflict develops between the sexes, it comes second only if the cooperation breaks down.

So which is first and which is second, conflict

followed sometimes by cooperation, or cooperation followed sometimes by conflict?

I favor the cooperation-then-conflict view because of why organisms reproduce sexually to begin with. From an evolutionary standpoint, reproducing sexually is an option, not a necessity. Animals are quite capable, as species, of reproducing asexually. Such species don't have any males, and the females produce eggs that don't need to be fertilized. Many animal species reproduce asexually, at least for a while, and many more could. If you are breeding chickens, for example, and you wait long enough, a female will arise having a mutation that leads her to make eggs that don't need to be fertilized and just hatch themselves. In fact, in the 1960s, turkeys and chickens were bred to make all-female strains, as described in professional journals at the time. So, why evolutionarily do females bother mating with males when they could in principle make fertile eggs that hatch by themselves?

The reason for sex is to share genes. The family tree of life shows quite clearly that asexual species go extinct quickly. Only the sexual species last through geologic time and grow new branches. Hence, it seems to me that the premise underlying all sexual reproduction is sharing—the sharing must come first and the conflict second.

The problem with sexual selection theory is not only the conceptual issue of which is first, cooperation or

conflict, in the male–female relationship. Sexual selection theory also doesn't square up with lots of facts
about sex roles in animals. The supposedly general
pattern of male–female behavior that Darwin wrote
about simply isn't as general as he apparently thought
it was.

In many species, the Darwinian sex-role pattern is
reversed, with the females flashy and the males drab,
the opposite of peacocks. Even schoolchildren know
that seahorse males have a little pouch that the female
deposits eggs into. The male seahorse incubates the
eggs and gives birth. Because females make eggs faster
than males can incubate them, females compete with
one another for eligible males to carry their eggs to
term. The females can be more colorful than the males,
and the males get to choose them, rather than the other
way around. Biologists call this "sex-role reversal."

In many species, moreover, the males and females
are equally "passionate" and choose each other with
equal care, called "partial sex-role reversal." And in
many species, like penguins, the males and females look
alike—they're hard to tell apart, and unlike peacocks,
neither sex is flashier than the other.

At the penguin extreme, all animals in the species
share one appearance. The other extreme is species
with multiple kinds of males and females. In many fish,
birds, and lizards, males and females come in two or
more kinds that differ in how long they live, whether

they defend territories, and whether they show off at the time of mating. In San Francisco Bay, for example, males of a fish called the plain-fin midshipman come in two types—one sings (yes, sings underwater) to females and has special nerves and musculature to do this, and the other type is silent. There are two kinds of coho salmon males, the "hooknose" and the "jack." And in many lizard species there are two or more kinds of females as well. For many species, no single standard describes the male and the female.

In addition, the extent to which the male–female binary itself breaks down is underappreciated. About 30 percent of the fish you see when snorkeling on a coral reef are members of species that feature sex change—that is, the same individual goes from making sperm to making eggs or vice versa, or makes both at the same time. Sex changing in animals is not some rare, bizarre thing. A stable male–female binary is a precondition for Darwin's discussion of sex roles, and even this binary can't be taken for granted.

Furthermore, there's no place in Darwin's theory for the same-sex mating that occurs in many species. Many think that homosexuality is rare in nature except in cases of mistaken identity. But homosexuality occurs often in nature. Penguins turn out to offer a rather humorous example of recent relevance.

The movie *March of the Penguins,* distributed by National Geographic Feature Films and Warner Independent Pictures, is widely known for its lovely photography of emperor penguins in the Antarctic. That photography and the publicity the movie has received in Christian circles have made it the second highest grossing documentary ever, according to the *New York Times* in September 2005.

Actor and screenwriter Jordan Roberts rewrote the original French narrative for American audiences using penguins to dramatize a "story of beauty," claiming "penguins have no agenda." The conservative columnist Maggie Gallagher enthused, "It is hard not to see the theological overtones in the movie he remade. Beauty, goodness, love and devotion are all part of nature, built into the DNA of the universe." The conservative film critic and radio host Michael Medved said it is "the motion picture this summer that most passionately affirms traditional norms like monogamy, sacrifice and child rearing." Rich Lowry, editor of the conservative *National Review,* told a gathering of young conservatives, "I have to say, penguins are the really ideal example of monogamy." Andrew Coffin, in the widely circulated Christian publication *World Magazine,* wrote that the survival of penguin eggs under the harsh conditions of life on an Antarctic ice pack makes "a strong case for intelligent design." And Richard Blake of the film studies program at Boston

College observed that the film was open to a religious interpretation: "I could see it as a statement on monogamy or condemnation of gay marriage."

Alas, the movie is not accurate. I don't know how the commentators missed it, but the country's major newspapers in the preceding months had been running front-page stories on gay penguins at New York and Japanese zoos. The *New York Times* reported in February 2004 that Roy and Silo, two chinstrap penguins at the Central Park Zoo in Manhattan, have been monogamous and "completely devoted" to each other for more than six years. "At one time, the two seemed so desperate to incubate an egg together that they put a rock in their nest and sat on it, keeping it warm in the folds of their abdomens," said their chief keeper, Rob Gramzay. Finally, he gave them a fertile egg that needed care to hatch. Things went perfectly, and a chick, Tango, was born. For the next two and a half months they raised Tango, keeping her warm and feeding her food from their beaks until she could go out into the world on her own.

The story of gay penguins began to take on a life of its own. In September 2005, Tom Musbach of PlanetOut.com went on to report the "devastating news" that Roy and Silo had broken up. After six years of a committed relationship, Silo had fallen for a female penguin named Scrappy. The "same-sex-household wrecker" had even begun building a nest with Silo, according to the *New York Post*. Penguins, along with other species in which homosexuality is present, including humans, generally practice homosexuality mixed in with heterosexuality.

It's now clear that homosexuality is a common feature of many animal social systems, just as it is in humans. In animals with backbones, like mammals, birds, reptiles, and fish, more than three hundred species have had homosexuality documented in the gold standard of professional journals in which articles can be published only after approval by a panel of anonymous scientific reviewers. The details differ widely across species. In some the homosexuality occurs only between females, in others only between males, and in some within both sexes. In some species about 10 percent of the matings are same-sex; in others as much as 50 percent or more, as in a kind of chimpanzee called the bonobo that, along with the common chimpanzee, is our closest relative. And the figure of three hundred species is conservative. So where does all homosexuality fit into Darwin's theory of universal sex roles for animals? It's a problem. According to Darwin, lots of species must be very confused.

Well, I think the factual case against the universality of the Darwinian sex roles is overwhelming and should be enough to falsify sexual selection theory on the spot. The factual difficulties with sexual selection theory have led to many ingenious work-around explanations.

Nonetheless, sexual selection theory's fundamental conceptual problem of getting the order of cooperation and conflict in male–female relations backward can't be evaded.

So what are we to do about this? How are we to explain the peacock's tail if Darwin's theory doesn't, or how do we explain homosexuality in animals, and all the other traits about how animals carry out reproductive social behavior where sexual selection has tried and failed? Well, I don't know. Replacing an entrenched theory like sexual selection will take a long time. To avoid the dissatisfaction of leaving the matter totally unresolved, let me toss out some of my own ideas about how the peacock's tail and homosexuality have evolved. Then we'll continue with gender and sexuality in the Bible.

I've never had the chance to study peacocks in the wild, but like many people, I've watched chickens. According to Darwin, the rooster has his cock-a-doodle-do, fleshy combs, and colorful feathers to impress females, to advertise his good genes so they will want to mate with him. But if you actually watch chickens, the hens couldn't care less about a rooster's showing off. A rooster is showing off to other males, not to females. Maybe the cock-a-doodle-do and the fleshy combs on a rooster, as well as the peacock's tail, are really intended for male–male interactions, not as advertisements to females. Maybe males check each other out to see if their ornaments are good enough to join them in power-holding cliques. These traits may be expensive admission tickets to exclusive all-male clubs that control access to who gets a chance to mate. Just an idea.

Turning to homosexuality, I believe this can be lumped with many other traits of physical intimacy among animals, including mutual grooming, preening, and synchronized back-and-forth calling. Perhaps these behaviors allow animals to coordinate their activities so that they can function together in teams. These behaviors enable teamwork and keep one another on the same page, similar to how athletes in team sports always touch one another, slap each other on the back, clap hands in a high five, and hold hands in a huddle, all to ensure coordination so that they can work as a unit.

Perhaps, too, animals even acquire, through their physical intimacy, a sense of real joy in friendship. Playing as a team needs not only coordinated action but also an ability to perceive the team welfare. The animals must be able to balance their sense of self-betterment with team betterment. A capability for team play can evolve because genes for this spread through the gene pool in successive generations— evolutionary success through cooperation rather than competition. I'm suggesting that cooperation in animals is coordinated play that furthers a team goal and is attained through bonds of friendship developed with behaviors involving physical intimacy.

Perhaps homosexuality in animals is a tip-off that we're overlooking how widespread cooperation is in nature. I don't dispute that evolutionary success depends on self-interest, but competition is not the same as self-interest. Competition is getting ahead regardless of the impact on others. If one insists that competition is the only way to succeed evolutionarily, then homosexuality seems a waste of time and likely some sort of mistake or disease. But physical intimacy, including homosexuality, is the way to come out ahead in the long run evolutionarily because of achieving effective team play and building bonds within the social infrastructure that allow offspring to be successfully reared.

Regardless of whether my pet theories for the peacock tail and homosexuality are correct, here's the situation today. The sex-role component of evolutionary theory implies that the male–female relationship is basically one of conflict that may, at best, be papered over with subsequent negotiation. This piece of evolutionary theory is challenged by its inconsistency with why sexual reproduction exists to begin with, which is to share genes, and by a great many cases of animal species that do not obey the standard sex roles that Darwin envisioned.

Gender and Sexuality

Although this discussion of gender and sexuality in animals may make for an interesting nature show, you may wonder if it has anything specifically to do with Christian faith. Darwinism views males and females as being in fundamental biological conflict. This implies that domestic violence is inevitable and suggests that living up to St. Paul's teaching that man and woman should become united as one body through marriage is impossible. Darwinism views marriage as a cover-up for never-ending conflict. In contrast, I view the male–female relationship as fundamentally one of cooperation that may degenerate into conflict but not necessarily. If so, St. Paul's ideal of intimate and total cooperation between a man and a woman in marriage becomes more attainable, although certainly not easy.

The data that refute Darwin's theory on sex roles come from the many species in which gender and sexuality are expressed in ways some Christians find morally suspect. This situation suggests that we reexamine the basis in scripture for regarding diversity in gender and sexuality as sinful to see if this dissonance may evaporate.

Many people assume the Bible unequivocally condemns homosexuality. In fact, only a very few biblical passages pertain to homosexuality, and their meaning has little to do with homosexuality as such. Perhaps the best known is a one-liner from Leviticus 18:22: "Thou shalt not lie with mankind, as with womankind: it is abomination." This passage has long been publicized as a direct prohibition against homosexuality. But is it?

Leviticus 18:22 speaks only of a class of sexual positions defined in antiquity as involving "penetration." Certain positions were reserved for male–female sex and others for male–male sex, including the "intercrural" position where the male couple stands face-to-face and one party thrusts between the thighs of the other. As the classicist K. J. Dover has shown, the ancient Greeks considered this position of male–male sexuality to be "clean" because it didn't involve what they defined as penetration.

Leviticus 18:22 merely polices allowable sexual positions. For a man to lie with a man as he would with a woman is an abomination, like eating seal, a creature of the sea without fins or scales. Instead, a man could lie with a man as only two men can lie with one another, such as in the intercrural position, and thereby follow the letter of the law. Because Leviticus is focused only on what positions to assume during sexual intercourse, it is irrelevant to the larger issue of including gays and lesbians in Christian ministry and of recognizing homosexual relationships as legitimate and sacred.

Similarly, the story of Sodom is often cited as condemning homosexuality. Again, it bears a closer look. Two angels approached the city of Sodom in disguise. Lot, a recent Bedouin immigrant to the city, extended his hospitality by inviting them into his home for the night. While Lot prepared food for them, news of their arrival spread across town. The men of Sodom surrounded Lot's house and "called unto Lot, and said unto him, Where are the men which came in to thee this night? Bring them out unto us, that we may know them." (Gen. 19:5) The angels blinded the men of Sodom so they could not continue their attack and instructed Lot and his family to leave the city immediately, and Sodom was then destroyed by an earthquake. The men of Sodom intended to rape two visitors, unaware that they happened to be male angels. It's the attempted rape itself, rather than its homosexuality, that is the sin.

The narrative of Sodom reveals a pattern in how the Bible mentions homosexuality. Homosexuality is never discussed as such, but only in combination with other issues like the violation of hospitality to visitors and rape that are the real focus of the teaching in Genesis. Many later references to Sodom refer to all manner of sins committed there. For example, the transgressions of Sodom are spelled out in Ezekiel 16:49–50: "Behold, this was the iniquity of thy sister Sodom, pride, fulness of bread, and abundance of idleness was in her and in her daughters, neither did she strengthen the hand of the poor and needy. And they were haughty, and committed abomination before me: therefore I took them away as I saw good." Pride, gluttony, idleness, not giving to the poor, as well as lewdness and fornication all comprise Sodom's sins.

The New Testament also records a long list of Sodom's sins: "And turning the cities of Sodom and Gomorrha into ashes ... making them an ensample unto those that after should live ungodly; ... chiefly them that walk after the flesh in the lust of uncleanness, and despise government. Presumptuous are they, selfwilled, they are not afraid to speak evil of dignities." (2 Peter 2:6, 10) While all of these actions merit disapproval, although I hope poking fun at politicians during the *Tonight Show* doesn't qualify as "speaking evil of dignities," it's wrong to single out homosexuality as the single sin of Sodom for which the city was destroyed.

St. Paul is the only disciple who explicitly mentions homosexuality at all. In his letter to the Romans, while condemning people who have fallen into worshiping human and animal images, St. Paul describes depravity that includes "lusts" and "vile affections." (Rom. 1:24, 26) The next two verses place "lust" within a long list of depraved behaviors such as "fornication, wickedness, covetousness, maliciousness" and condemns those who are "haters of God, despiteful, proud, boasters." (Rom. 1:29, 30) The people who have abandoned God suffer all these evils collectively. St. Paul's letter does not focus on any particular vice for special condemnation but holds up the whole suite of behaviors as symptomatic of losing touch with God.

St. Paul then remarks about the orgies of those who have lost their way, and homosexuality appears here explicitly. Included among the many vices in such orgies are "women did change the natural use into that which is against nature . . . And likewise also the men, leaving the natural use of the woman, burned in their lust one toward another; men with men working that which is unseemly." (Rom. 1:26–27) St. Paul's teaching is not about homosexuality as such, but the overall wanton behavior that is symptomatic of losing faith in God's love. St. Paul's letter guides his followers back to God; it does not hold forth on sexual orientation.

Most important, Jesus himself does not condemn or even mention homosexuality. The scarcity of passages that refer to homosexuality raises the question of why the Bible seems so silent on the issue of sexual orientation. The Bible doesn't condemn homosexuality, but the Bible doesn't directly endorse it either—there are no passages about baptizing homosexual couples or clear sentences saying that homosexual people have a place in the church. The Bible doesn't pay any attention to homosexuality as such, one way or the other.

What are we to make of this silence about homosexuality? Perhaps we're not looking in the right place. When I was a child in Sunday school, passages about eunuchs were mentioned briefly and quickly dismissed. Eunuch was an archaic category, not relevant to our modern world, limited to children castrated to sing soprano in choirs. In fact, the category of eunuch is much larger than boy sopranos.

The Bible mentions eunuchs in many contexts. A search for the keyword "eunuch" on a King James Bible Web site returns thirty verses over ten books. Most important, the great prophet Isaiah and even Jesus himself speak at length of eunuchs—not in one-liners or offhand phrases but in entire parables.

In the Old Testament, Isaiah teaches, "For thus saith the Lord unto the eunuchs . . . unto them will I give in mine house and within my walls a place and a name." (Isa. 56:4, 5)

In the New Testament, Jesus describes three kinds of eunuchs: those "which were so born from their mother's womb," those "which were made eunuchs of men," and those "which have made themselves eunuchs for the kingdom of heaven's sake." (Matt. 19:12) Jesus' characterization of eunuchs matches the Roman description that lumps people who were intersexed (born eunuchs); who were castrated, often as an act of war (man-made eunuchs); and who voluntarily became eunuchs on their own into a single all-encompassing category. The eunuch category included gender-variant people like priestesses to the goddess Cybele and others who switched gender, usually under religious auspices. Jesus does not describe eunuchs further but continues with his inclusive message, "Suffer little children, and forbid them not, to come unto me;

for of such is the kingdom of heaven." (Matt. 19:14) Because of the proximity of the verses, we surmise that Jesus would have welcomed eunuchs, too.

An explicit instruction to include eunuchs within the church appears later, in Acts, where baptism is detailed. The apostle Philip met a "eunuch of great authority" who was returning from Jerusalem, where he had gone to worship. He was sitting in his chariot reading Isaiah. The spirit told Philip to approach the chariot, and the eunuch invited Philip to travel with him. "And as they went on their way, they came unto a certain water: and the eunuch said, See, here is water; what doth hinder me to be baptized? And Philip said, If thou believest with all thine heart, thou mayest. And he answered and said, I believe that Jesus Christ is the Son of God. And he commanded the chariot to stand still: and they went down both into the water, both Philip and the eunuch; and he baptized him." (Acts 8:27–38)

No one disputes the Bible's extensive endorsement of eunuchs in both the Old and New Testaments, but one might argue that the ancient category of eunuchs is obsolete and has nothing to do with the gender- and sexuality-variant people of our times. In fact, as the historian Mathew Kuefler has shown, the Roman description of eunuchs spans all our contemporary categories. Some eunuchs were feminine-identified. Firmicus Maternus reported, somewhat disparagingly, on eunuchs who "feminized their faces, rubbed smooth their skin, and disgraced their manly sex by donning women's regalia ... They nurse their tresses and pretty them up woman-fashion; they dress in soft garments." Apuleius said that such eunuchs renounced their previous masculine identities and called one another "girls" in private. Such eunuchs were evidently marrying as women, too. Other eunuchs were boyish and partook of homosexual relations with older men. And still other eunuchs were successful in the public space of men, holding powerful positions as ministers in imperial court and leading military campaigns. Eunuchs were common enough that writers referred to them with phrases such as "a crowd of eunuchs, young and old," "armies of eunuchs," "troops of eunuchs," and so forth.

The human sexual diversity that we see in today's America has always existed, on the streets of ancient Jerusalem and Rome as now in San Francisco, New York, Atlanta, or any American city. These are the very people to whom Isaiah offers "a place and a name," whom Philip affirms "mayest" receive baptism, and whom Jesus accepts as having "made themselves eunuchs for the kingdom of heaven's sake."

We can now answer the question of why the Bible seems silent about homosexuality. Homosexuality as a category of personal identity emerged relatively recently—during the late 1800s in Europe. Before that, homosexuality occurred of course but wasn't part of one's personal identification any more than one's preference for various colors or flavors are today. The issue of including homosexuals would not have been raised then any more than would the issue of including people today depending on whether they preferred grays or primaries for their colors or bitter or sweet in their beverages. Today we carve up the spectrum of human diversity into different descriptive categories than the ancients did. Yet the underlying spectrum of human diversity is still evident in the ancient world, and the Bible's affirmation of all those people is relevant to us now.

So, I wind up with a hopeful vision of unity between a revised understanding of family, gender, and sexuality in both evolutionary biology and Christian teaching. It seems to me the facts of nature support a much more optimistic view of whether approaching the ideal of cooperation between man and woman in marriage is possible than standard sex-role theory in biology advises, and passages of the Bible support a much more inclusive welcome to persons of varying gender expression and sexuality than some denominations currently take as policy. Still, each of us will have to ponder these points in our own time and way. It's a fact of our times that these issues have come to the forefront, for whatever reason. Certainly, biologists are uncomfortable and dismayed at the prospect of revising or even discarding sex-role theory from our discipline's master texts on evolution, and some denominations will not appreciate a suggestion to reconsider policies condemning gender- and sexuality-variant people that have often been in place for centuries. So change, if it comes, will come only slowly, as each of us in our own time and way, by ourselves and in communion with one another, decides what is right.

12

Future Directions

THINK THE DISPUTATIOUS atmosphere surrounding the teaching of evolution in our schools distracts us from more important issues that would benefit from Christian moral perspective. How can we move beyond the disputation to restore the courtesy that marked the courtly exchanges between Charles Darwin, an agnostic, and the American botanist Asa Gray, a deeply religious Christian, during the late 1800s?

I suggest we first identify positions that needlessly provoke polarization and learn to avoid them. Then, each of us, one by one, and in groups and community, can insist on a new standard of discourse. This may be followed by shifting some of the energy now being dissipated on anti-evolution advocacy into other issues.

In my opinion, these issues include reforming the narratives of selfishness that underwrites social Darwinism, developing nuanced moral positions about genetic engineering and cloning, and improving our care for the global dimensions of God's creation.

Here are the positions to avoid, one associated with evolutionary biology and the other with Christian televangelism.

In *The Selfish Gene,* the evolutionary biologist Richard Dawkins publicizes a view of nature emphasizing competition: "We are survival machines—robot vehicles blindly programmed to preserve the selfish molecules known as genes." He continues, "Our genes made us. We animals exist for their preservation and are nothing more than their throwaway survival machines. The world of the selfish gene is one of savage competition, ruthless exploitation, and deceit." In *River Out of Eden,* Dawkins writes, "The universe we observe has precisely the properties we should expect if there is, at bottom, no design, no purpose, no evil and no good, nothing but blind pitiless indifference." And in *Devil's Chaplain,* he says, "Blindness to suffering is an inherent consequence of natural selection. Nature is neither kind nor cruel but indifferent." He also writes in *The Extended Phenotype* (a biological term for the traits that genes produce) that the gene should be thought of as having "its effects on the world at large, not just its effects on the individual body in which it happens to be sitting."

These four books develop a philosophy of universal selfishness as though that were a fundamental part of evolutionary biology. This philosophy not only goes far beyond the data of evolutionary biology but is incorrect as well. It fails to appreciate the difficulty of disentangling the contribution of individual genes to the whole and about how a whole can function in ways beyond the sum of its parts.

Take cooking as an example. When you add salt to water you get salt water. You can then boil the water off, collect the vapor, cool it, and re-collect the water, with the salt remaining at the bottom. In this way you can reconstitute both ingredients. Salt water is no more than the sum of salt plus water.

If, however, you add flour to water and heat, you get a form of bread. There's nothing you can do to reconstitute the flour and water. You've made a new compound that assumes a function and significance beyond the sum of flour and water. It has acquired its own identity and function because of the chemical *bonds* that have formed between water and flour molecules.

Similarly, when genes combine to make a body, the body becomes a unit more than the sum of the genes in it because the body now functions as a unit. And when athletes play together as a team, they acquire an identity and perform differently than the sum of individual players. That's what teamwork is all about, the difference between a doubles game of tennis and two singles games.

An interview with Dawkins in *The Guardian* on February 10, 2003, illustrates the problem of the selfish-gene philosophy. During the interview, Dawkins "wanders over to the other side of the room and returns with a bird's nest that he picked up in Africa. 'It's clearly a biological object.' His eyes light up. 'It's clearly an adaptation. It's a lovely thing.' He says that birds do not need to be taught to make nests, they are genetically programmed to do so."

When Dawkins refers to a bird's nest, the issue I wish to raise is not whether a bird's ability to make a nest is genetically endowed, but whether one or two birds cooperated to make it. A male bower bird makes a showy nest by himself that he uses in courtship, and such a nest can be thought of as an extension of his body. In contrast, a male and a female together make a robin's nest, and their nest for holding eggs is not the extension of the body of either bird, but a new entity that results from the bond between the male and the female robin, from their collaboration. This nest, which is a fruit of their relationship, is what allows them to breed and to raise young together. The evolutionary success of a male and a female robin resides not in the genes they have as individuals but in the relationship they develop with each other. The conceptual problems that result from overemphasizing individualism, and from the difficulties of defining what an "individual" is, run throughout selfish-gene philosophy.

Most evolutionary biologists are inclined to dismiss the selfish-gene metaphor as an entertaining hyperbole. For our present purposes, however, the matter can't be left there because selfish-gene proponents directly and continually attack religion. In an interview in *Slate* magazine on April 28, 2005, Dawkins stated, "Religion is scarcely distinguishable from childhood delusions like the 'imaginary friend' and the bogeyman under the bed. Unfortunately, the God delusion possesses adults, and not just a minority of unfortunates in an asylum." Dawkins further claims that religion is a kind of mental virus, or "meme," an idea that readily replicates within easily infected minds. Dawkins wrote in *Devil's Chaplin,* "To describe religions as mind viruses is sometimes interpreted as contemptuous or even hostile. It is both." In fact, Dawkins is so outspoken in his opposition to religion that the Los Angeles–based, politically progressive though anti-religious Atheist Alliance instituted the Richard Dawkins Award in 2003.

The rhetoric of these attacks provokes reprisals that are equally harsh and in turn strengthen political support for anti-evolutionists. Because evolutionary biologists as a group have not sufficiently distanced themselves from selfish-gene philosophy, some parents and community groups oppose teaching evolution in the schools out of fear that the teaching will not be limited to the actual science of evolution but will also present an ideology of selfishness as though that were science, too.

More important, selfish-gene philosophy intellectually enables the intelligent design movement. If one group (selfish-gene advocates) asserts that facts from science refute the existence of God, then another group (intelligent design advocates) is free to conclude the opposite with as much validity. Selfish-gene advocates can't have it both ways—they can't assert that science disproves the existence of God and then turn around and say that anyone with the opposite position isn't doing science. Inasmuch as most scientists think the existence of God can't be proved with data, the nonexistence of God can't either.

Let's now examine quotations from someone on the other "side," someone as opposed to teaching evolution as Dawkins is to practicing religion.

The 700 Club, a television program seen by about one million Americans a day, is the flagship program of the Christian Broadcasting Network founded by Pat Robertson, a U.S. Christian televangelist. Over the years, Robertson has frequently been in the news. In an interview on the September 13, 2001, telecast of *The 700 Club*, the conservative Christian evangelist Jerry

Falwell characterized the September 11 tragedy as payback from an angry god, and Robertson agreed. Falwell stated, "What we saw on Tuesday, as terrible as it is, could be miniscule if . . . God continues to lift the curtain and allow the enemies of America to give us probably what we deserve . . . I really believe that the pagans, and the abortionists, and the feminists, and the gays and the lesbians who are actively trying to make that an alternative lifestyle, the ACLU, People For the American Way—all of them who have tried to secularize America—I point the finger in their face and say 'you helped this happen.'" To which Robertson replied, "Well, I totally concur."

More recently, Robertson has also weighed in about teaching evolution. In the November 2005 elections, all eight members of the now-infamous Dover, Pennsylvania, school board who were up for reelection were defeated. In response, Pat Robertson warned residents that disaster may strike them because they "voted God out of your city." On his *700 Club* broadcast, Robertson said, "I'd like to say to the good citizens of Dover: If there is a disaster in your area, don't turn to God. You just rejected him from your city." Later, Robertson added, "If they have future problems in Dover, I recommend they call on Charles Darwin. Maybe he can help them."

The narrative of nature as universally selfish and deceitful and the narrative of a hating, wrathful God mark opposite poles in today's thought. Selfish-gene philosophy distorts evolutionary biology, and the wrathful-God theology distorts Christian teaching. Just as the tales of the selfish gene energize opposition to teaching evolution, threats of God's vengeance energize opposition to religion. We simply don't have to let ourselves get caught up in these polarizing positions. We can insist on a better tenor of discourse.

How do we get a better tenor of discourse, apart from not listening to polarizing advocacy? We can forgive. If someone who has uttered inflammatory statements about the selfishness of nature or the vindictiveness of God's wrath wishes to return to the center, we should forgive and not rub his or her face in the past. We should welcome those who wish a peaceful dialogue, regardless of their history of dispute. It may be hard for people who are now polarized not to think they are letting their side down by moving to the center, and it may be hard for those smitten by the rhetoric to forgive, but this is what making peace takes. And I believe we do need to make peace so that we can move on to other issues. If we can accept the facts and theory of evolution as "settled science," we can redirect our energies to some serious moral issues we now face. So let me indicate three moral issues that I think are the most important and what I think we can do about them.

I suggest we begin by disputing the secular philosophy of our day that glorifies competition, the dog-eat-dog survival of the fittest, as excusable and even meritorious because such conduct supposedly expresses basic human nature. This is the philosophy of "social Darwinism," which I think amounts to the application of a mistaken understanding of evolution to human social behavior. We shouldn't allow opposition to social Darwinism and emphasizing cooperation in its place to be dismissed as romantic, wishy-washy thinking. Philosophers refute social Darwinism on the grounds that "science says what is, not what ought to be." Therefore, even if evolution were a nasty business, it shouldn't matter for human ethics. That's nice, but this dismissal doesn't end the matter. I'm troubled by how the science itself is continually misrepresented. Nature is not simply dog-eat-dog survival of the fittest, and therefore the possibility of transferring this view of nature from animals to humans is incorrect to begin with. So what can be done? We need to understand and to publicize better the biology of animals with complex social systems in which organisms do not live as simple individuals but as members of social groups.

How can we ordinary people promote this? On a personal level, we can encourage our friends and relatives who work or study in biology. Let me give you a sense of how the personal touch can make a difference. Most of us biologists spend holidays with family when we can. Here's what it's usually like when visiting at

Thanksgiving. A relative asks, "What are you doing these days?" If you say, "I'm working on the Portuguese man-of-war," a colonial jellyfish whose components need to cooperate, you know what you're going to hear: "Ugh, I hate those slimy things." I can't tell you how much it would mean if a relative said instead, "That's wonderful. I hear this may change our whole way of seeing nature." You don't have to be that gushy. Any word of encouragement—any—at Thanksgiving dinners in thousands of homes across the country could turn what is often an ordeal for young biologists into an inspiring encounter. Encouraging people whose work emphasizes social cooperation and deemphasizes individualism would do more to change the secular culture of universal selfishness than any number of lawsuits about teaching evolution ever could.

Another area to highlight is the moral dimension of biotechnology. The media during December 2005 reported a spectacular fraud in stem cell research committed by Hwang Woo Suk of Seoul University in South Korea. *Time* magazine had named Hwang one of the People Who Mattered in 2004, describing him as "the first to clone human embryos capable of yielding viable stem cells that might one day cure countless diseases . . . Hwang has already proved that human cloning is no longer science fiction, but a fact of life." It turns out that Hwang fabricated the results of papers published in 2004 and 2005 that reported success at human cloning. I'm personally relieved that human cloning is still science fiction, because this gives us a breather to think about the morality not only of cloning but of other aspects of biotechnology as well.

I don't think that human cloning for medical purposes is coming soon. In *Evolution's Rainbow,* I predicted that cloning would be slow to develop because I was suspicious of the theory behind it. So when Hwang's papers appeared in 2004, I assumed I was wrong. But it now seems worthwhile to repeat what I believe is wrong about the idea of cloning—a problem similar to the selfish-gene premise.

Cloning begins by extracting an egg from the ovary of an adult female. An egg consists of the nucleus, where most of the genes are located, and the surrounding material called cytoplasm. The nucleus from the female's egg is destroyed, say, by irradiation, supposedly leaving the cytoplasm intact. Next, the cell to be cloned is fused (merged) with the egg cell that is missing its nucleus. The nucleus from the cloned cell is then supposed to take over the merged cytoplasm and direct all subsequent activities.

This narrative is one of genetic control—an active nucleus inserted into passive cytoplasm. It would seem, however, that the state of the cytoplasm has a role of its own to play in whether the nucleus can carry out the instructions contained in its genes. If so, a cell's functioning reflects the *relationship* between the nucleus and the cytoplasm. The failure to consider both the cytoplasm and the nucleus as co-equal in development explains, I conjecture, why cloning is not working well. The practical problem with cloning is that it takes a great many eggs, about 250 or more, which means as many as 250 women have to submit to having eggs extracted from their ovaries, to get one cytoplasm–nucleus combination that happens to function together. Cloning a cell is a shot in the dark. This is because, I think, the cytoplasm isn't prepared to work with just any old nucleus plopped into it.

The undoubting and total commitment of molecular biologists to their narrative of genetic control explains why such a major fraud has occurred in stem cell research. Scientific fraud is nearly impossible to detect when people believe the anticipated results before the experiment is actually done. Then when results are reported, everyone says, "of course" and congratulates the researchers.

Most molecular biologists believe cloning will work. Every week scientists are quoted as saying so. For example, when interviewed by the *New York Times* at his Harvard Stem Cell Institute on January 24, 2006, Douglas Melton said, "Even though Hwang's findings turned out to be fraudulent, nothing he claimed was a fundamental challenge to the principles of embryonic stem cell research. Hwang claimed to have successfully cloned human embryos, which was only a technical accomplishment. It's something that's already been done in animals." This confidence that cloning is only a soon-to-be-solved technical problem sets up the research culture described above in which fraud becomes nearly undetectable.

Hwang claimed not merely to have cloned humans, but to have made the procedure very efficient, so that about 5 eggs were needed instead of 250 to get a working clone cell. This claim is fraudulent, which means that cloning remains inefficient. I'm questioning whether the continued inefficiency of cloning, plus the failure of investigators in other countries to do any better than Hwang, signifies that cloning principles are really incorrect after all because the nucleus–cytoplasm relationship has not been taken sufficiently into account.

My guess is that we still have several years before molecular biologists realize the need to shift their attention away from the nucleus as the controller of the cell and to focus instead on the nucleus–cytoplasm relationship. When that happens, say, three or four years from now, cloning will become a reality. In the meantime, we should quickly get ready for the future and assess the status of our moral thought about cloning and related technologies.

Christian moral discussion of cloning is typified by the position of the Roman Catholic Church. In many publications the Church has linked cloning of embryonic cells to the issue of abortion, to the "destruction of developing human life" as Father Tadeusz Pacholczyk refers to it in a brochure published by Family Research Council (FRC). Writings of Father Pacholczyk, who holds a Ph.D. in neuroscience from Yale, together with those of Richard Doerflinger of the U.S. Conference of Catholic Bishops, bring a healthy skepticism toward claims of "therapeutic benefits" from the biotechnology industry.

However, I'd like to invite Christians to consider reframing the moral issues surrounding cloning somewhat more broadly. In the FRC brochure, Fr. Pacholczyk raises the question, "Isn't it a matter of religious belief as to when human beings begin?" He answers with, "It is not a matter of religious belief, but a matter of biology . . . It is a matter of empirical observation. Once you are constituted as a human being (which always occurs at fertilization or at an event that mimics fertilization like cloning), you are a new member of the human race."

But the biology is not as clear-cut as Fr. Pacholczyk suggests. When the sperm enters the egg in human fertilization, the nuclei of egg and sperm do not fuse until the two-cell stage, and the sperm genes don't "turn on" right away either. At its beginning the embryo is more a "confederation" than a union. Furthermore, the embryo can't develop into a baby until it implants into the uterus, which occurs about six days after

fertilization. Until then the embryo is still accumulat-
ing the pieces it needs, including the placenta (the
afterbirth connected to the umbilical cord). The placenta surrounding the baby is made from the baby's
cells together with the mother's cells and is a shared
structure. The exact point between fertilization and
implantation when the embryo should be considered
as human, when it acquires its soul, can't be resolved as
a matter of biology. Recall that even Pope John Paul II
said of the "sciences of observation" that "The moment
of transition into the spiritual cannot be the object of
this kind of observation."

Let's consider other issues instead. In assisted-
reproduction technologies such as artificial insemina-
tion, the egg and sperm are united in a laboratory and
the embryo is then placed in a woman so it may implant
in her uterus, develop a placenta with her, and eventually be born as a baby. This technology also permits the
embryos to be screened while still in the laboratory and
particular ones to be selected for implantation. Pre-
implantation screening leads to what *Time* magazine
once termed "designer babies." Manufacturing designer
babies amounts to breeding people. The technique is
advertised as a way to avoid implanting embryos carry-
ing genetic defects, but actually it could, and almost
certainly would, be used for screening for any trait the
parents wished—or any trait an employer or government wished.

In addition, the stage prior to implantation is when genes can be added to the embryo. Genes are added by using a "friendly virus," a normally pathogenic virus whose harmful capability has been deactivated. A virus can carry a gene from one organism to another. Viruses have been used to transfer genes to make genetically engineered plants that are resistant to herbicides and even for frivolous activities such as putting a gene for green fluorescence from a jellyfish into a rabbit as a form of "living art." These are not science fiction fantasies. As long ago as 2002, the *Wall Street Journal* reported that human genes had been introduced into cows, sheep, goats, rabbits, and mice so that these animals would grow human biochemicals or body parts that could be isolated or harvested for transplantation into humans. Thus, it is perfectly possible to introduce all manner of genes into human embryos while they are in the pre-implantation stage. None of these procedures involves cloning because an egg and a sperm are combined as usual to produce an embryo. It's the next step that raises the deep moral issues: What's being done once the embryo has been made? How many human genes does a pig need to have before it qualifies for human rights, or at least some rights? Is a pig with a human kidney just a pig? And how many animal genes does a person need to have before losing his or her human rights? If we can lift our eyes from arguments about whether the soul enters the embryo at fertilization or six days after fertilization, we can then see on the horizon much deeper moral issues that affect the definition of humanity itself.

This is not the place to try to resolve the moral questions raised by today's biotechnology. But it is the place to start raising the questions and ask how prepared we are to answer them. I'd love to discuss the moral issues of biotechnology within a community of faith. But most church congregations and their leaders are not prepared for those discussions. I know some denominations are holding seminars and retreats to brief clergy on the issues, and even the basic vocabulary, of biotechnology. This is a wonderful start. I believe each of us in the pews should join in discussing the moral aspects of biotechnology. The university forums I'm aware of seem loaded on behalf of biotechnology, and bioethicists seem mostly interested in finding ways to excuse biotechnology. I don't ever hear the bioethics community saying "No, this is wrong" about biotechnology. But as Christians we *will* say that something is right and something else is wrong. We are now called to develop nuanced moral stands about what is right and wrong with these new technologies in whatever way balances the health benefits to those already alive with protection for the sanctity of humanity.

The final moral issue I wish to highlight is our Christian responsibility to care for God's creation. God created the nonliving and the living parts of the earth to work together. In Genesis 2, the Bible states that after the seventh day when God rested, he created the earth processes on which life depends, the wind and the rain. Before then, "These are the generations of . . . every plant of the field before it was in the earth, and every herb of the field before it grew: for the Lord God had not caused it to rain upon the earth, and there was not a man to till the ground. But there went up a mist from the earth, and watered the whole face of the ground." (Gen. 2:4–6) Thereafter, God created Adam from "the dust of the ground" (Gen. 2:7) and the plants "out of the ground" (Gen. 2:9). We now know that the common material origin of Adam and of the plants from the ground is manifested in all of life's being united as one gigantic family tree. We now also know how intricately the life that God created is intertwined with the "mist of the earth" and the waters on the face of the ground.

I've worked for many years not only on lizards but on seashore animals—the creatures you see on rocks and in tide pools at the ocean's edge. These animals, such as starfish, barnacles, mussels, snails, crabs, lobsters, and even the tiny fish that live in rock crevices, all release their eggs to the waters. When the eggs hatch, the tiny young are called larvae, and they float in the waters of the ocean eating tiny one-celled plants that float along with them. They depend on winds and currents to eventually bring them back to the shore where they were born so that they may complete their life cycle and produce young of their own. If the winds fail, the young animals die at sea, and people living at the ocean's edge can wonder why there are no crabs this year, or lobsters, or salmon. For the life cycle of many animals to go around, the breath of God must continue to blow across the waters.

It may seem that only animals of the sea depend so intimately on earth processes and that we humans are far removed from such dependency. Consider, then, Moses. To protect him from threat of death, his mother "took for him an ark of bulrushes, and daubed it with slime and with pitch, and put the child therein; and she laid it in the flags by the river's brink . . . And the daughter of Pharaoh came down to wash herself at the river . . . and when she saw the ark among the flags, she sent her maid to fetch it. And when she had opened it, she saw the child . . . And she had compassion . . . And she called his name Moses: and she said, Because I drew him out of the water." (Exodus 2:3–10) Every adult starfish, barnacle, mussel, snail, crab, lobster, and tiny fish living in a rock crevice is drawn from the water as Moses was.

Similarly, without the wind, there would be no corn-fields in Kansas, because the wind carries corn's pollen from flower to flower. The wind turns the life cycle of many plants on land just as water turns the life cycle of animals in the sea. Our call to care for God's living creation therefore extends to stewardship of the winds and the waters.

As to the plants and animals themselves, God charged Adam with naming creation: "whatsoever Adam called every living creature, that was the name thereof." (Gen. 2:19) What would Adam do today? The birds have been named. The cattle have been named. Most vertebrates are now named. What of "every creeping thing that creepeth upon the earth" that also disembarked with Noah from the ark? (Gen. 8:17) Do we say, "Adam, you're too late. The trees of your garden have been felled, and the animals with them are lost"? Yes, we must say this, because we have extinguished so many species from the planet. This is not how to care for God's creation.

I'm overjoyed at news reports that the National Association of Evangelicals (NAE), which consists of 45,000 churches with 30 million parishioners, is considering adopting a policy statement supporting care for the global environment. As reported in the *Financial Times* on December 23, 2005, the NAE together with the Southern Baptist Convention is studying the "Sandy Cove Covenant," which includes the following statement of principles: "We invite our brothers and sisters in Christ to engage with us the most pressing environmental questions of our day, such as health threats to families and the unborn, the negative effects of environmental degradation on the poor, God's endangered creatures, and the important current debate about human-induced climate change." It continues, "We covenant together to engage the evangelical community in a discussion of climate change with a goal of reaching a consensus statement on the subject in 12 months." This is a beautiful and inspired statement about one of the major moral issues facing Christians today—how to care for God's creation in view of the sheer magnitude of human activity on the planet.

I thank you for taking the time to read this book. Although we have considered some differences of opinion among Christians and explored paths for reconciliation, I wish to close with Jesus' teaching that sums up our common faith:

> *Thou shalt love the Lord thy God with all thy heart, and with all thy soul, and with all thy mind. This is the first and great commandment. And the second is like unto it, Thou shalt love thy neighbour as thyself. On these two commandments hang all the law and the prophets.* (Matt. 22:37–40)

Index

❋

Index

Index

Index

Index

Index

Index

Index

Index

About Island Press

Island Press is the only nonprofit organization in the United States whose principal purpose is the publication of books on environmental issues and natural resource management. We provide solutions-oriented information to professionals, public officials, business and community leaders, and concerned citizens who are shaping responses to environmental problems.

In 2006, Island Press celebrates its twenty-first anniversary as the leading provider of timely and practical books that take a multidisciplinary approach to critical environmental concerns. Our growing list of titles reflects our commitment to bringing the best of an expanding body of literature to the environmental community throughout North America and the world.

Support for Island Press is provided by the Agua Fund, The Geraldine R. Dodge Foundation, Doris Duke Charitable Foundation, The William and Flora Hewlett Foundation, Kendeda Sustainability Fund of the Tides Foundation, Forrest C. Lattner Foundation, The Henry Luce Foundation, The John D. and Catherine T. MacArthur Foundation, The Marisla Foundation, The Andrew W. Mellon Foundation, Gordon and Betty Moore Foundation, The Curtis and Edith Munson Foundation, Oak Foundation, The Overbrook Foundation, The David and Lucile Packard Foundation, The Winslow Foundation, and other generous donors.

The opinions expressed in this book are those of the author(s) and do not necessarily reflect the views of these foundations.